Laser Refractography

B. S. Rinkevichyus • O. A. Evtikhieva
I. L. Raskovskaya

Laser Refractography

 Springer

Prof. B. S. Rinkevichyus
Dept. Physics
Moscow Power Engineering Institute
Krasnokazarmennaya 17
111250 Moscow, Russia
bronyus@yandex.ru

Dr. I. L. Raskovskaya
Dept. Physics
Moscow Power Engineering Institute
Krasnokazarmennaya 17
111250 Moscow, Russia
raskovskayail@mpei.ru

Dr. O. A. Evtikhieva
Dept. Physics
Moscow Power Engineering Institute
Krasnokazarmennaya 17
111250 Moscow, Russia
evtikhievaoa@mpei.ru

ISBN 978-1-4419-7396-2   e-ISBN 978-1-4419-7397-9
DOI 10.1007/978-1-4419-7397-9
Springer New York Dordrecht Heidelberg London

Library of Congress Control Number: 2010937790

Printed on acid-free paper

Springer is part of Springer Science+Business Media (www.springer.com)

Preface

This monograph is devoted to the description of the physical fundamentals of **laser refractography**—a novel informational-measuring technique for the diagnostics of optically inhomogeneous media and flows, based on the idea of using spatially structured probe laser radiation in combination with its digital recording and computer techniques for the differential processing of refraction patterns.

Considered are the physical fundamentals of this technique, actual optical schemes, methods of processing refraction patterns, and possible applications. This informational technique can be employed in such areas of science and technology as require remote nonperturbative monitoring of optical, thermophysical, chemical, aerohydrodynamic, and manufacturing processes.

The monograph can also be recommended for students and postgraduates of informational, laser, electro-optical, thermophysical, chemical, and other specialties.

Laser refractography is a conceptually novel refraction method for the diagnostics of inhomogeneous media, based on the idea of using spatially structured probe laser radiation in combination with its digital recording and computer techniques for the differential processing of refraction patterns.

Structured laser radiation (SLR) in laser refractography can be treated as a spatial, two-dimensional discrete multiposition signal. The *refractogram* (image) of SLR passing through the inhomogeneous medium under study is recorded in the plane of observation with a multiposition receiver and compared, following its preliminary processing, with standard images corresponding to various inhomogeneity models. This approach makes it possible to apply the methods used for the correlation analysis of discrete spatial multiposition signals to refractograms, these being essentially the same. The results of comparison between experimental and standard refractograms, made on the basis of correlation or some other metric, are used to select such inhomogeneity model as fits best the experimental refractogram.

By varying the combination, orientation, and arrangement of the elementary components of SLR one can *adapt* the measuring system to the structure of the inhomogeneity in hand, the extended character of the radiation source providing for the *simultaneity* of the diagnostics of the process in different regions. The *discrete* character of SLR suits perfectly the up-to-date digital image recording and processing techniques, which makes possible high-precision *quantitative diagnostics* of the medium being studied.

Laser refractography came into existence owing to great achievements in laser and information technologies, such as the creation of diffraction optical elements (DOE) and development of efficient digital techniques for the recording and processing of optical images.

By virtue of the practically *inertialess* character of refraction measurements, laser refractography can be used to diagnose not only stationary, but also *nonstationary fast processes*. What is more, thanks to the possibility of producing narrow probe beams, the method is suitable for the diagnostics of *boundary layers* and *edge effects*, as well as investigations into processes taking place within micro- and nanochannels.

Laser refractography can first of all be employed in such areas of science and technology as require remote nonperturbative monitoring of thermophysical, chemical, aerohydrodynamic, and manufacturing processes. For this reason, the present monograph will prove useful to specialists engaged in the development of laser diagnostic techniques and their application in the research practice of their laboratories. It can also be recommended for students and postgraduates of these specialties.

This monograph generalizes the many years of practical experience of the authors and the research staff of V. A. Fabrikant Chair of Physics of Moscow Power Engineering Institute, with due regard for the most important results achieved by scientists in the Russian Federation and other countries.

The authors thank Professor V. A. Grechikhin, Head of Chair of Fundamentals of Radio Engineering of Moscow Power Engineering Institute (Technical University), for his assistance in writing Chap. 7, A. V. Tolkachev, Leading Research Worker of V. A. Fabrikant Chair of Physics, for the preparation and implementation of original experimental investigations, and Assistant Professor N. M. Skornyakova for her help in the computerized typesetting of the manuscript. The authors express their sincere gratitude to their coworkers M. V. Esin, V. A. Orlov, E. V. Savchenko, and E. V. Pudovikov, whose results were partially included into this monograph, and also to K. M. Lipitsky, A. V. Mikhalev, V. T. Nguen, and E. V. Stepanov, postgraduate students of the Chair of Physics, for their assistance in doing the computer graphics.

Contents

About the Authors

Bronyus Simovich Rinkevichyus graduated from Moscow Power Engineering Institute with a degree in applied physical optics in 1965, defended his candidate's thesis in electronics at Moscow Power Engineering Institute in 1969, defended his doctoral thesis in optics at P. I. Lebedev Physical Institute of the Russian Academy of Sciences in 1980, Full Professor at V. A. Fabrikant Chair of Physics of Moscow Power Engineering Institute (Technical University), Doctor of Physics and Mathematics Science, author and coauthor of over 300 scientific papers, 3 monographs, and 5 books published in Russian and English, Member (affiliate) of IEEE, Full Member of D. S. Rozhdestvensky Optical Society, Full Member of the International Academy of Sciences of Higher School. The main fields of his scientific interests are coherent and informational optics, physical fundamentals of the laser diagnostics of flows, and history of science.

Olga Anatolyevna Evtikhieva graduated from Moscow Power Engineering Institute with a degree in optoelectronic instruments in 1975, defended her candidate's thesis at Moscow Power Engineering Institute in 1980, Head of V. A. Fabrikant Chair of Physics of Moscow Power Engineering Institute (Technical University), Candidate of Technical Sciences, author and coauthor of over 80 scientific papers and 4 books. The main fields of her scientific interest are applied and informational optics and laser refractometry.

 **Irina Lvovna Raskovskaya** graduated from Moscow Power Engineering Institute with a degree in radio physics in 1981, graduated from Moscow State University with a degree in theoretical physics in 1990, defended her candidate's thesis at Moscow State Technical University Stankin in 2005, Senior Researcher at V. A. Fabrikant Chair of Physics of Moscow Power Engineering Institute (Technical University), Candidate of Physics and Mathematics Science, author and coauthor of over 50 scientific papers. The main fields of her scientific interests are propagation of radio and optical waves in inhomogeneous media and laser measuring systems.

Conclusion

This monograph expounds the physical principles of laser refractography—a novel informational-computer laser technique for the diagnostics of optically inhomogeneous media and flows, based on the idea of using spatially structured laser radiation in combination with its digital recording and computer techniques for the differential processing of refraction patterns.

To sum up, it can be said that the three whales whereon laser refractography rests are:

- Structured laser radiation
- Digital methods for recording and processing refraction patterns
- Special software for the solution of the inverse problem of inhomogeneity profile reconstruction

The joint use of these laser refractography constituents offers qualitatively new possibilities for investigations into various optically transparent media, which bespeaks the creation of an innovative informational technique for the diagnostics of optically inhomogeneous media and flows.

The fundamentally new capabilities of this technique include:

- Adaptation to the shape of the surface under study
- Formation of informative three-dimensional images
- Simultaneous measurements in different spatial regions (which is important in the studies of nonstationary processes)
- Quantitative diagnostics of inhomogeneities
- Direct informative visualization during the course of monitoring of the processes of interest

In addition, the laser refractography technique possesses all the merits inherent in laser measurement methods, such as remotability, practically zero lag, and capabilities of taking nonperturbative and microscopic measurements.

As follows from the above-described procedure of measuring temperature in boundary layers, an important virtue of laser refractography is its capability of *quantitative diagnostics* of the parameters of the media under study and reconstruction of inhomogeneity profiles on the basis of comparison between experimental and theoretical refractograms. The error of this method is mainly determined by diffraction effects and can substantially be reduced through the use of special computer image processing techniques.

Laser refractography can be used to monitor stationary and nonstationary fast processes, including thermal processes in liquids, gases, and plasmas, natural convection processes in liquids near heated or cooled bodies, and processes of mixing of different liquids in process vessels, and also to quantitatively diagnose temperature fields in boundary layers, acoustic fields, and fields of the other physical quantities that influence the refractive index of the media under study.

In our further investigations, we intend revealing the capabilities of the laser refractography technique to record and monitor such fundamental phenomena as the origination of the singular points of convective instability, edge effects at the borders of solids, formation of micro- and nanostructures under special heating conditions, and so on.

Chapter 1
Introduction

1.1 Optical Refraction Methods for the Diagnostics of Various Media

The recent active application of laser methods for the diagnostics of acoustic pressure, temperature, density, salinity, and current velocity fields in transparent media [1] is due to their substantial advantages over other methods. First, optical measurements do not disturb the fields under study, for the energy absorbed by the medium being probed is in most cases sufficiently low. Moreover, laser methods are practically devoid of inertial errors, which makes it possible to diagnose fast processes. Their additional merit is the possibility they provide of taking remote measurements. Laser methods allow one to investigate a refractory index field that can then be converted into the desired field of another physical quantity.

The present-day stage of development of laser and computer techniques is characterized by the advent of visible semiconductor lasers and DOE [2], digital video and photocameras capable of resolving over a million of picture elements (pixels), and computers with an operating speed in excess of 3 GHz and a memory capacity over 100 GB, along with the development of novel efficient digital methods for processing optical images.

Refraction techniques for the diagnostics of inhomogeneous media and flows have in recent years gone through renewal conditioned by the factors mentioned above. Matrix photoreceivers and computers have made it possible to develop novel schemes for laser gradient refractometers, namely, speckle refractometry, computer-laser refraction method (COLAR) [3], and background-oriented schlieren (BOS) method [4]. High-quality laser beams differing in shape have been produced with simple optical elements, and scanning and multichannel refractometric systems have been developed [5].

Laser refractography (LR) is a novel laser method for the diagnostics of optically inhomogeneous media, based on the phenomenon of refraction of structured laser radiation (SLR) in optically inhomogeneous media, digital recording of the refraction patterns (refractograms), and their computer processing. This method is

B. S. Rinkevichyus et al. (eds.), *Laser Refractography,*
DOI 10.1007/978-1-4419-7397-9_1, © Springer Science+Business Media, LLC 2010

being used for the visualization and quantitative investigation of transparent stationary and nonstationary inhomogeneous media.

Before proceeding to the presentation of the fundamentals of this method, its
merits and shortcomings, we should briefly consider the principles of the classical methods of visualization of inhomogeneous media. Several basic visualization
methods exist in optics, namely, phase-contrast, schlieren, interference, polarization, and holographic ones. The first two, which bear direct relationship to the subject matter of this monograph, are classed with gradient refraction methods based
on the refraction of optical radiation in media featuring a gradient refraction index.
And laser refractography should also be placed into the same class.

The evolutionary history of gradient refraction methods [6–8], since Robert
Hooke's time and till the present day, can arbitrarily be divided into two stages: the
classical stage, where use was made of wide collimated optical beams and spatial
filtration of refraction patterns, and the laser-computer stage, employing structured
laser radiation and computer processing of refraction patterns. Let us present in this
aspect a description and classification of the main gradient refraction techniques,
determine the place of laser refractography within the scope of this classification,
and compare it with other methods (see Fig. 1.1).

As follows from Fig. 1.1, the basis of the given classification is the division of
the visualization methods according to the character of the optical radiation they use.

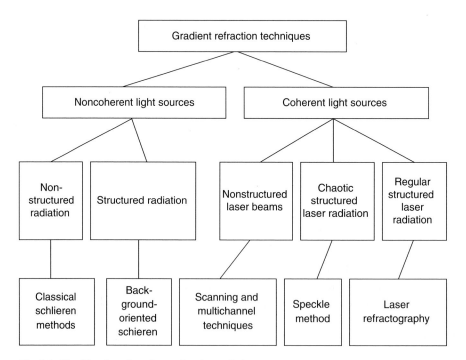

Fig. 1.1 Classification of gradient refraction techniques

The *classical schlieren methods* [6–8] (see Sect. 1.2) predominantly use non-coherent light sources and wide, practically homogeneous optical beams—a characteristic example of nonstructured radiation. Radiation featuring distinct spatial continuous or discrete intensity modulation of chaotic or regular character will be considered to be structured.

In the *background-oriented schlieren* (BOS) method (see Sect. 1.3), the originally homogeneous noncoherent radiation from the source becomes modulated in intensity while reflecting from a chaotically or regularly structured screen against whose background the transparent inhomogeneity being visualized is observed. The inhomogeneity is visualized as a result of the refraction-induced displacement of the structural elements of the screen.

The *scanning and multichannel refractometric systems* (see Sect. 1.4) use narrow laser or collimated noncoherent beams, and the refraction-induced displacement of the beams serves as a measure of the integral index of refraction; that is, the method here is quantitative. A single scanning beam should preferably be classed with nonstructured radiation, but a set of beams in a multichannel laser refractographic system can be treated as a prototype of discretely structured laser radiation.

In the *speckle method* (see Sect. 1.5), coherent laser radiation passing through a diffuser forms a chaotic speckle structure against whose background the inhomogeneity being visualized is observed. Visualization occurs owing to the refraction-induced displacement of the structural speckle elements. As regards the way of observing the object, the speckle method is similar to the background-oriented schlieren technique.

In contrast to the speckle method, *laser refractography* (see Sect. 1.6) employs a regularly structured laser radiation formed on the basis of special optical elements directly at the exit from the radiation source. This method of forming SLR provides for its high coherence and small beam divergence, which makes it amenable to description in terms of geometrical optics. In these terms, SLR can be modeled by families of beams generating surfaces in the form of a discrete set of planes, nested cylinders, cones, and so on.

Drawing an analogy with the BOS method, one can treat the source of SLR as an active structured screen. But in contrast to BOS, the high intensity and directivity of SLR require that the object visualization method be changed, and so SLR passing through the inhomogeneity of interest is projected onto a screen in the plane of observation to form the so-called 2D refractogram. The high intensity of SLR makes it possible to observe by scattered light 3D refractograms, i.e., surfaces formed by geometric-optical beams undergoing refraction.

The discrete and regular character of refractograms makes them perfectly suitable for digital recording and differential computer processing, which in contrast with the optical processing of images forming the basis of the schlieren methods allows attaining high precision in the quantitative diagnostics of inhomogeneity profiles.

For one to have a more vivid notion of the methods being classified, we describe in the sections below some ways of implementing them.

1.2   Classical Schlieren Methods

Zernike Phase-Contrast Method [9–11]. The optical scheme of a setup to implement the Zernike method is presented in Fig. 1.2. Radiation from point light source
1 passes through lens *2* forming plane wave front *3* and illuminates phase object *4*
(an optical inhomogeneity) that transforms the plane wave front into deformed front
5. The visualization of this front is effected with lens *6* and spatial filter *7* placed
in the focal plane of lens *6*. The refraction image is recorded in plane *8* where an
intensity distribution is formed that linearly depends, under certain conditions, on
the phase distribution in the deformed wave front.

Image recording plane *8* is conjugate to the plane wherein phase object *4* lies;
i.e., the distances l_1 and l_2 and the focal distance f satisfy the lens formula

$$1/l_1 + 1/l_2 = 1/f.$$

The crucial point in the given method is the spatial Zernike filter placed in the focal
plane of the objective lens. The filter is a small disk-shaped plate whose thickness is

$$d = \lambda_0/4(n-1),$$

where λ_0 is the wavelength of light in vacuum and n is the refraction index of the
plate material.

The Zernike phase plate produces a phase shift of $\Delta\phi = \pi/2$ for the central diffraction maximum. If w_0 is the phase plate radius in terms of spatial frequencies, the
amplitude transmission coefficient of the Zernike filter will be

$$G(w) = 1, \quad \text{if } |w| > w_0, \quad \text{and} \quad G(w) = -j, \quad \text{if } |w| < w_0.$$

If the phase object causes small phase shifts $\varphi(x, y) \ll 1$, the amplitude transmission
coefficient of such an object may be written down in the form

$$t(x,y) = \exp\{ j\varphi(x,y)\} \approx 1 + j\varphi(x,y).$$

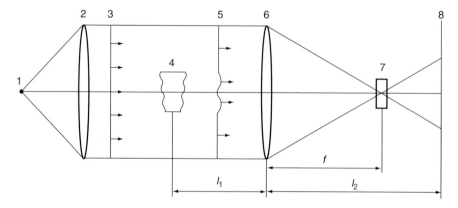

Fig. 1.2 Optical scheme of visualization of a phase object by the Zernike method

Behind the phase plate the amplitude transmission coefficient is given by

$$t'(x,y) \approx j + j\varphi(x,y).$$

For this reason, the light intensity distribution in the image plane will be

$$I(x,y) \approx I_0[1 - 2\varphi(x,y)], \tag{1.1}$$

where I_0 is the light intensity in the image plane in the absence of the phase object. One can see from the above expression that the intensity contrast of the filtered image is linearly related to the phase distribution function of the object being studied and is a small quantity. One can enhance contrast by reducing the amplitude of the zero-order diffraction maximum τ times, so that expression (1.1) will assume the form

$$I(x,y) \approx I_0[1 - 2\tau\varphi(x,y)]/\tau^2.$$

One should pay attention to the fact that the linear relationship between $I(x, y)$ and $\varphi(x, y)$ occurs only if $\varphi(x, y) \ll 1$, which substantially restricts the applicability of the given method to the analysis of phase objects.

The Zernike method has found widespread application in the microscopy of biological objects that in most cases belong in the class of phase objects. Fritz Zernike won the Nobel Prize for physics in 1953 for the development of this method and the invention of the phase-contrast microscope.

Schlieren Methods Based on Amplitude Filters. Experimental aerodynamics makes wide use of schlieren methods based on the Foucault knife-edge placed in the focal plane of the image-forming lens [6]. The optical scheme of the setup implementing such a method is similar to that presented in Fig. 1.1, but the Zernike phase plate here is replaced by an amplitude filter—the Foucault knife-edge. Various types of knife-edges placed at various angles to the horizontal axis are used in practice. Figure 1.3 illustrates three positions of Foucault knife-edges used to visualize the refraction index gradient of a medium in various directions.

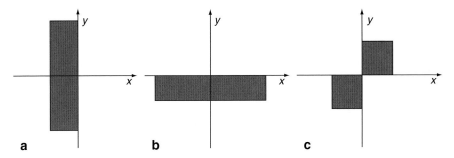

Fig. 1.3 Arrangement of Foucault knife-edges: **a** to visualize the gradient of n along the x-axis; **b** to visualize the gradient of n along the y-axis; **c** two-dimensional Foucault knife-edge to visualize the gradient of n along the x- and y-axes simultaneously

Fig. 1.4 Visualization of a shock wave in a supersonic gas flow

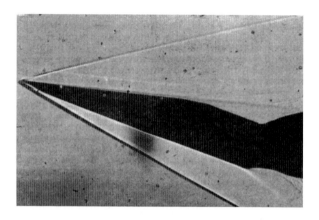

Figure 1.4 shows as an example the visualization of a shock wave in a supersonic flow past a wing.

Instead of the Foucault knife-edge that divides the spatial frequency plane into two regions with light transmission coefficients of $\tau_1 = 1$ and $\tau_2 = 0$, one can use an *absorption filter with three gradations* [10] that divides this plane into three regions, namely, a dark region with $\tau_1 = 0$, gray region with $0 < \tau_2 < 1$, and light one with $\tau_3 = 1$. The width of the gray region located on the optical axis corresponds to that of the zero-order diffraction maximum. This filter makes it possible not only to visualize wave-front phase gradients differing in magnitude, but also to differentiate between them by their sense.

Schlieren Methods Based on Phase Filters. A phase filter in the form of a phase knife is placed in the spatial frequency plane so that the edge of the step producing a phase shift of $\Delta\varphi = 180°$ passes through the point where $w_x = w_y = 0$ and is oriented along either the *x*- or *y*-axis. The analysis of the operation of such a filter by Hilbert-optics methods [10] shows that in contrast to the situation with the Foucault knife-edge there is no persistent background light here. The sensitivity of schlieren instruments with phase knives is many times (around 5) as high as that of their counterparts using the amplitude Foucault knife-edges. Use can also be made of phase filters in the form of diffraction gratings with an irregularity [11].

Defocusing Methods. A special place in the schlieren instruments is occupied by the visualization method based on the defocusing of the refraction pattern; i.e., the placing of the image recording plane outside of the conjugation region [10], and also the deflectometric methods based on two Ronchi gratings [12]. Another version of the defocusing method is based on the Talbot effect, i.e., the self-reproduction of the image of a periodically transmitting transparency upon diffraction of light.

A distinctive feature of these methods is the fact that the refraction image obtained is determined by the second derivative of the phase function. If a phase object having a periodic structure is illuminated with a collimated light beam, images of the object appear in image planes spaced from the object at distances of

$$z_p = 2md^2/\lambda;$$

m is an integer and d is the period of the structure.

To visualize such an object, one should carry out observations in one of the visualization planes spaced from the object at distances of

$$z_v = (m + 1/2)(d^2/\lambda).$$

Figure 1.5 presents a visualization scheme using two Ronchi gratings (moire deflectometry technique) between which the phase object under study is placed.

Thus, the above brief analysis of the classical schlieren methods shows that:

- all the methods considered above make use of radiation with a plane or spherical wave front;
- to visualize phase objects, use is made of the optical filtration of refraction images in the focal plane;
- the ideal Zernike filter visualizes the phase function (x, y) without any distortion;
- the Foucault knife-edge and the phase knife transform the amplitude transmission coefficient of the object into its Hilbert transform; the absorption filter with three gradations visualizes the phase function gradients $(d/dx)\ \phi(x, y)$ and $(d/dy)\ \phi(x, y)$;
- the defocusing and moire deflectometry methods visualize the second derivatives of the phase function, $(d^2/dx^2)\ \varphi(x, y)$ and $(d^2/dy^2)\ \varphi(x, y)$.

Table 1.1 lists comparative characteristics of the most widespread schlieren methods.

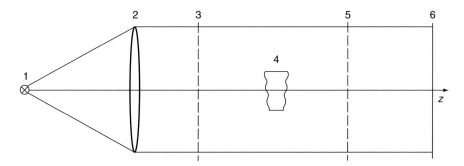

Fig. 1.5 Optical scheme of visualization of an object by the moire deflectometry method: *1*—point light source, *2*—objective lens, *3* and *5*—Ronchi grids, *4*—object under study, *6*—image recording plane

Table 1.1 Comparative characteristics of schlieren methods

Nos.	Method	Filter form	Visualization characteristics
1	Zernike phase-contrast method		$\varphi(x,y)$
2	Foucault knife edge		$\|d\varphi(x, y)/dx\|$, $\|d\varphi(x, y)/dy\|$
3	Phase knife; Hilbert filter		$\|d\varphi(x, y)/dx\|$
4	Filter with three gradations		$\pm d\varphi(x, y)/dx$
5	Defocusing; moire deflectometry		$\|d^2\varphi(x, y)/dx^2\|$

1.3   Background-Oriented Schlieren Method

The present-day stage of development of schlieren methods has been marked by the suggestion that use should be made of structured screens illuminated with noncoherent radiation [4, 13–15]. The method suggested can easily be implemented by means of either a structured transparency or a structured reflecting screen illuminated with a noncoherent point source, e.g., an incandescent lamp. The transmission or reflection coefficient here can be either a deterministic or a random coordinate function.

A schematic diagram of a setup implementing this method is presented in Fig. 1.6. It is made up of noncoherent light source *1* that uniformly illuminates semitransparent structured transparency (screen) *2*. The image of this transparency is projected by the objective lens of digital camera *4* on to CCD array *5*. Phase object *3* under study is placed between screen *2* and the objective lens of digital camera *4*.

This is an imaging refractometric scheme; i.e., the objective lens produces an image of the transparency (screen): the distances l_1 and l_2 and the focal distance f satisfy the lens formula $1/l_1 + 1/l_2 = 1/f$.

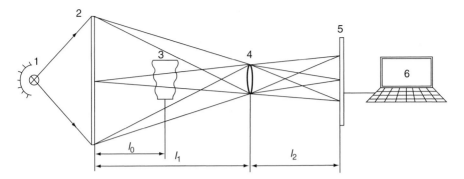

Fig. 1.6 Optical scheme of a refractometric setup using noncoherent structured radiation: *1*—noncoherent light source, *2*—semitransparent structured transparency (screen), *3*—phase object, *4*—objective lens of digital camera, *5*—CCD array, *6*—personal computer

The structured transparency comprises a set of dark dots or figures of some other shape arranged in a deterministic or random fashion on a light ground, or light dots or other figures on a dark ground. Two images of the transparency (screen) are recorded, one in the absence of the phase object and the other in its presence. Computerized comparison made between them with the aid of special software enables one to judge of the main characteristics of the phase object. This comparison yields a picture showing the magnitude of the refraction index gradient averaged over the light propagation path. To solve the inverse problem also proves possible in some cases.

Figure 1.7 presents typical forms of structured screens, and Fig. 1.8 the sequence of steps in obtaining a refraction image of a hot ball immersed in cold water [15].

The manufacture of transparencies and screens of arbitrary structure presents no great difficulties. Natural formations, such as grass, sand, snow, asphalt, can also be used as a screen. Large-size screens can also be produced, which makes this method suitable for studying extensive phase objects, up to a hundred meters long. To investigate such objects by the classical methods is impossible. Refraction patterns can be processed by both standard [16] and special-purpose [15] optical image processing techniques.

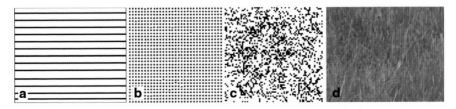

Fig. 1.7 Typical forms of structured screens. **a** Deterministic arrangement of lines. **b** Deterministic arrangement of dots. **c** Random arrangement of dots. **d** Natural screen—grass

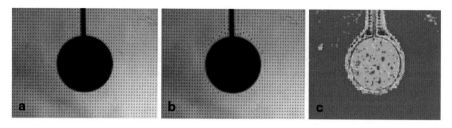

Fig. 1.8 Sequence of steps in obtaining a refraction image of a hot ball immersed in cold water. **a** Screen image with the ball whose temperature is equal to that of the surrounding medium. **b** Screen image with the hot ball. **c** Computer processing result

The transverse spatial resolution R_t is governed by the angle of view of the objective lens; i.e., the distance l_1 from the screen to the objective lens, the distance l_0 from the screen to the phase object, and the diameter D of the objective lens (see Fig. 1.6):

$$R_t = l_1/Dl_0,$$

whence it follows that as far as the improvement of the resolution R_h is concerned, it is necessary to reduce the distance l_0; i.e., bring the screen closer to the phase object, and reduce the aperture of the objective lens. This means that the depth of focus of the objective lens should be increased. Unfortunately, reducing the aperture of the objective lens reduces the light flux incident upon the CCD array as well.

Like any direct shadow technique, this method has a shortcoming that it is impossible to match together a sharp image of the transparency (screen), which is necessary for image processing, and a sharp image of the object being studied, which is required for the visualization to be informative.

It should be noted that subject to recording in this method is the displacement of the elements of structured radiation, rather than the variation of the illuminance in the image recording plane as is the case with the classical schlieren methods.

Klinge and Kompenhans [17] demonstrated the capabilities of this method in their studies into flows past helicopter propeller blades and aircraft wings.

1.4 Laser-Computer Scanning and Multichannel Methods

One of the principal disadvantages of the classical refraction methods is the fact that the information about the phase object of interest is contained in the illuminance distribution in the image recording plane. With laser-computer methods, this information is embodied in the displacement of isolated structural elements in the plane of observation; and the accuracy of measurement of displacement is much higher than that of the magnitude of illuminance.

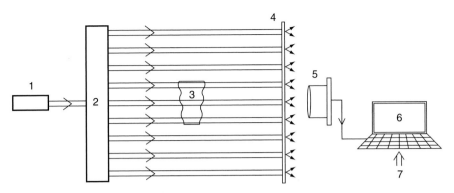

Fig. 1.9 Generalized schematic diagram of a laser-computer refraction system: *1*—coherent light source, *2*—structured radiation forming optical system, *3*—phase object, *4*—diffuse scattering screen, *5*—digital refraction image recording system, *6*—personal computer, *7*—software

Of course, the picture visualizing the phase object in a classical schlieren instrument can be photographed with a digital camera as well, but the processing of the image captured in the presence of the object cannot provide any additional information in comparison with the original image obtained in its absence. Another situation occurs with experiments by laser-computer methods that implement the differential principle of gathering information and use well-developed up-to-date methods of processing optical images, the cross-correlation technique for one. Moreover, digital image recording allows direct computer processing of images, which raises the refraction techniques to a new level, making them fit for quantitative diagnostics.

Figure 1.9 presents a generalized schematic diagram of a laser-computer refraction system. It consists of coherent light source *1*, structured radiation forming optical system *2*, diffuse scattering screen *4*, digital refraction image recording system *5*, and personal computer *6*, complete with software *7*. Phase object *3* under study is placed between system *2* and screen *4*.

The given refraction system serves to determine the characteristics of transparent optically inhomogeneous objects (media), optical glasses, crystals, liquids, gases, and plasmas.

Scanning Laser Refractometer [18–20]. The first laser refractometric investigations were conducted with precisely such beams, and their deflection from the rectilinear propagation path in optically inhomogeneous media was recorded with a position-sensitive photodetector.

A four-quadrant position-sensitive detector (PSD) makes it possible to measure the deflection of the laser beam in any direction, and it is not only the magnitude of the deflection that is determined, but also its direction. A merit of this measurement system is that its output signal is electrical, which allows using well-developed methods for its processing. The advantages of the system become apparent in studies of nonstationary processes, especially fast ones, for the time constant of PSDs is shorter than 10^{-6} s. Such a simple system was used to study temperature distribution in a boundary liquid layer near a hot vertical plate [19].

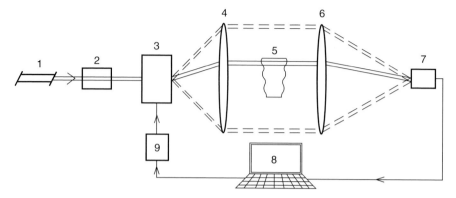

Fig. 1.10 Schematic diagram of a laser scanning refractometer: *1*—gas laser, *2*—optical system, *3*—scanner, *4* and *6*—objective lenses, *5*—phase object, *7*—PSD, *8*—personal computer, *9*—scanner control unit

Figure 1.10 shows the optical scheme of a laser scanning refractometer that makes it possible to investigate the structure of optical inhomogeneities by scanning them with a laser beam in one or two directions. The refractometer consists of gas laser *1*, optical system *2* forming a beam with desired characteristics, scanner *3*, scanner control unit *9*, objective lenses *4* and *6*, with phase object *5* under study between them, PSD *7*, and personal computer *8* that controls the operation of the scanner and processes the information obtained.

Gumennik and Rinkevichyus [19] described a combined system, built around a laser scanning refractometer and a Model ИАБ-458 commercial schlieren instrument, that was used to study the degeneration of turbulence in a stratified liquid [1].

Multichannel Refractometric System [21, 22]. The next step in the development of gradient refractometry using structured laser radiation involved the design of a multichannel refractometric instrument consisting of a system for forming a multiple beam and a PSD array (Fig. 1.11). This made it possible to investigate

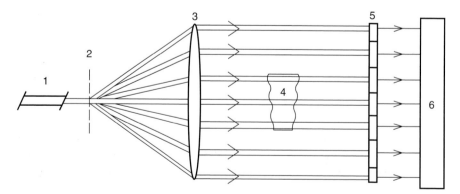

Fig. 1.11 Optical scheme of a multichannel laser instrument for measuring refractive index gradients: *1*—laser; *2*—2D diffraction grating; *3*—objective lens; *4*—phase object; *5*—PSD array; *6*—electronic unit

nonstationary gradient refraction index fields in optically inhomogeneous media. As compared with the classical schlieren instruments, this system offered substantial advantages as to operating speed. Its principal shortcoming was the small number of channels, depending on the number of elements in the PSD array. A comparative analysis of the classical, scanning, and multichannel laser gradient refractometric systems can be found in the work by Gumennik and coauthors [5].

The situation changed substantially with the advent of multichannel CCD photo-receivers that made it possible to design multichannel refractometric systems with up to 5×10^6 channels.

1.5 Speckle Refractometry

Structured laser radiation is a spatially amplitude-modulated light beam. Such beams are obtained from a plane or homocentric laser beam by placing in its path either classical optical elements like amplitude and phase filters, Ronchi gratings, plane-parallel glass plates, etc. or special DOE produced by computer optics methods.

The spatial modulation of a laser beam can be either deterministic or random (noise-like) with a modulation degree (contrast) close to unity. Also possible is a space–time modulation of the laser beam, for example, modulation of the beam in space through its scanning.

Consider one of the ways to produce chaotically structured radiation. Once the first gas lasers made their appearance in the sixties [23], a special property of their radiation was noted, namely, the granularity of beams reflected from rough surfaces. It later turned out that this granularity was due to the high spatial coherence of laser radiation, and it was given the name *speckle structure*. Actually, this is a typical chaotically structured radiation that is produced every time a laser beam passes through a diffuser or reflects from a diffuse-reflecting surface. A vast body of literature was devoted to the characteristics of the speckle structure of laser radiation; the monograph by Francon [24] may be recommended for first acquaintance with the matter.

Let us consider the main characteristics of speckle fields formed on the passage of laser radiation through a diffuser (Fig. 1.12).

Here a beam of high spatial coherence from gas laser *1* passes through negative lens *2*, and divergent beam *3* formed by the lens illuminates transparent diffuser *4* after which the laser radiation acquires a speckle, i.e., granular structure. This speckle structure (pattern) is observed with objective lens *6* in image recording plane *7*. The random light field distribution in plane *7* is independent of the properties of the diffuser and is solely determined by the wavelength of the laser radiation and the angle of view of objective lens *6*, that is,

$$d_{\rm s} \approx \lambda/\alpha \approx \lambda l/D,$$

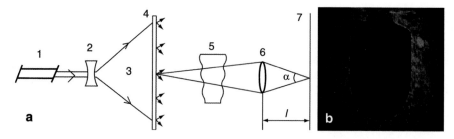

Fig. 1.12 Formation of a speckle structure upon coherent illumination of a diffuser. **a** Optical scheme: *1*—gas laser, *2*—negative lens, *3*—divergent beam, *4*—transparent diffuser, *5*—phase object, *6*—objective lens, *7*—image recording plane. **b** Photo of speckle structure

where d_s is the speckle diameter, α is the angle of view of the objective lens, l is the distance from the objective lens to the image recording plane, and D is the diameter of the objective lens.

Thus, the size of speckles can easily be changed by moving the objective lens or the image recording plane. The speckle structure wholly depends on the coherent properties of laser radiation and occurs as a result of interference of a multitude of scattered waves of random initial phase. A specific feature of the speckle structure is its high contrast, for in the scattered field there always exist points where the total field amplitude is close to zero. Shatokhina and coworkers [25] analyzed the scheme of an optical gradient refractometer relying for its operation on the properties of the speckle structure.

1.6 Laser Refractography

The essence of laser refractography is to probe the medium of interest with structured laser radiation, record the radiation passing through the medium with a digital camera, and process with the aid of a computer the refraction patterns captured with a view to finding out the properties of the medium.

To get a more graphic notion of the laser refractography technique, consider the main principles of its implementation.

Laser refractography was preceded by the method based on the use of plane-structured laser radiation (laser plane). This computer-laser refraction (COLAR) method of studying spatially inhomogeneous flows uses digital processing of refraction patterns obtained by probing the flow with one or several laser planes instead of a single wide light beam, as is the case with the classical schlieren methods [5]. Depending on the phenomenon being investigated, laser planes can be oriented in space arbitrarily. A digital video camera is used here to record changes in the shape of the laser-plane images observed on a semitransparent screen, as compared to their original condition.

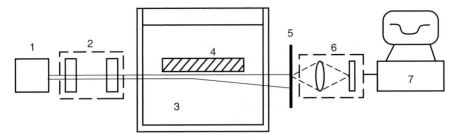

Fig. 1.13 Schematic diagram of an experimental setup implementing the COLAR method: *1*—laser, *2*—optical system, *3*—liquid-filled cell, *4*—hot body, *5*—diffuse-scattering screen, *6*—digital video camera, *7*—personal computer

Figure 1.13 presents a schematic diagram of an experimental setup implementing the COLAR method to study natural convection in the vicinity of hot bodies immersed in a liquid. It consists of laser *1*, optical system *2* for forming the laser plane, liquid-filled cell *3*, hot body *4*, diffuse-scattering screen *5*, digital video camera *6*, and personal computer *7*, complete with special software.

Figure 1.14 shows typical laser-plane images obtained with this setup.

Figure 1.15a shows a schematic diagram of a COLAR setup with crossed laser planes [26]. It is made up of He–Ne laser *1*, argon laser *2*, optical system *3* for forming the vertical laser plane, optical system *4* for forming the horizontal laser plane, beam splitting cube *5*, tilting prism *6*, flow *7* under study, semitransparent screen *8*, digital camera *9*, and personal computer *10*, complete with special software *11*. Figure 1.15b and c illustrate visualization of an inhomogeneous turbulent flow by means of such a system.

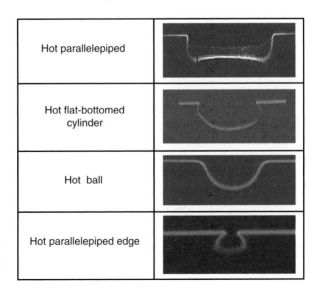

Fig. 1.14 Typical images of laser planes passing near hot bodies

Hot parallelepiped	
Hot flat-bottomed cylinder	
Hot ball	
Hot parallelepiped edge	

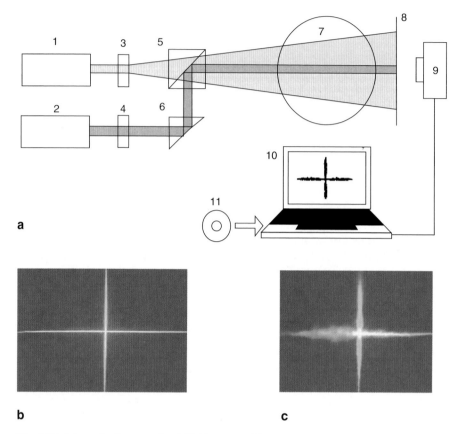

Fig. 1.15 Schematic diagram of a COLAR setup with crossed laser planes. **a** Optical scheme: *1*—He–Ne laser, *2*—argon laser, *3*—optical system for forming the vertical laser plane, *4*—optical system for forming the horizontal laser plane, *5*—beam splitting cube, *6*—tilting prism, *7*—flow under study, *8*—semitransparent screen, *9*—digital camera, *10*—personal computer, *11*—special software. **b** LP refractogram in a homogeneous flow. **c** LP refractogram in an inhomogeneous turbulent flow

The classical optical elements can help obtain only a very limited range of laser planes, namely, a narrow beam, plane, and cross.

The situation substantially changed with the advent of compact diffraction optical elements produced by computer optics techniques [2]. It was precisely this circumstance, combined with the present-day achievements in the field of semiconductor lasers, computer technology, and digital image processing techniques [16], that gave the onset to the novel line of laser gradient refractometry, known as laser refractography.

Figure 1.16 presents refractograms of a boundary layer near a hot ball in a water, obtained using structured laser radiation in the form of a set of conical beams produced by means of DOE.

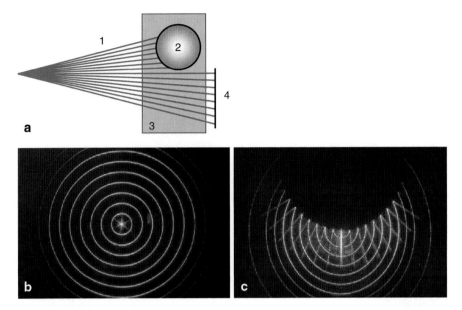

Fig. 1.16 Refractograms of a boundary layer near a hot ball, obtained using structured laser radiation in the form of a set of conical beams. **a** Experimental scheme: *1*—conical beams, *2*—ball, *3*—water-filled cell, *4*—screen. **b** Refractogram in the absence of the ball. **c** Refractogram in the presence of the hot ball in cold water

The "graphical" character of refraction SLR patterns suits the term *laser refractography*, and the refraction patterns themselves are called *refractograms*.

In essence, the 2D refractogram observed on the screen is a 2D image of the spatially structured radiation source that is formed by the optical system constituted by the medium under study.

A 3D *refractogram* is a 3D image of a surface formed by beams from a SLR source that undergo refraction in the medium of interest, and it can be obtained through processing by special methods a set of experimental or theoretical 2D refractograms from different sections of the medium. As stated earlier, a 3D refractogram can be visualized experimentally by scattered radiation. The use of SLR also makes it possible to effect 3D *visualization of caustic surfaces* and trace their formation on the basis of evolution of 2D refractograms.

A refractogram is, in a sense, a "portrait" of the medium in hand that can be compared with a set of elementary refractograms of typical inhomogeneities, which enables one to carry out, directly in the course of visualization, express diagnosis of the medium; i.e., to draw qualitative conclusions as to the variation of its parameters with time.

The use of digital refractogram recording and processing techniques, as well as special software to solve the inverse inhomogeneity profile reconstruction problem, allows one to conduct quantitative diagnostics of the inhomogeneity simultaneously with its visualization. It can, therefore, be said that laser refractography is a quantitative visualization technique.

By varying the combination, orientation, and arrangement of the elementary SLR components, one can adapt the experimental setup to the structure of the given inhomogeneity, the extended character of the radiation source providing for the simultaneity of diagnostics in different regions of the medium.

By virtue of the practically inertialess nature of refraction measurements, laser refractography can be used to study not only stationary, but also nonstationary, fast processes, including thermal processes in liquids, gases, and plasmas, natural convection in liquids near hot or cold bodies, quantitative diagnostics of temperature fields in boundary layers during the course of heating and cooling processes, liquid mixing processes in chemical manufacturing apparatus, and so on.

Moreover, thanks to the possibility of producing narrow probe beams, laser refractography is fit for studies into boundary and edge effects, as well as processes taking place in microstructures.

References

1. B. S. Rinkevichius, *Laser Diagnostics in Fluid Mechanics* (Begell House Inc. Publishers, New York, 1998).
2. V. A. Soifer, Ed., *Computer Optics Methods* (Fizmatlit, Moscow, 2003) [in Russian].
3. O. A. Evtikhieva, A. I. Imshenetsky, B. S. Rinkevichyus, and A. V. Tolkachev, "Computer-Laser Refraction Methods for Studying Optically Inhomogeneous Flows," *Izmeritelnaya Tekhnika*, 6, 15–18 (2004).
4. N. M. Skornyakova, E. M. Popova, B. S. Rinkevichyus, and A. V. Tolkachev, "The Investigation of Heat Transfer by Background Oriented Schlieren Method," in CD-ROM *Proceedings of the 12th International Symposium on Application of Laser Techniques to Fluid Mechanics, Lisbon, 2004.*
5. E. V. Gumennik, B. S. Rinkevichyus, O. A. Evtikhieva, and Y. D. Chashechkin, "Concurrent Use of Qualitative and Quantitative Refractometric Methods," *Inzhenerno-Fizichesky Zhurnal*, **50**(4), 597–604 (1986).
6. A. F. Belozerov, *Optical Methods of Visualization of Gas Flows* (Kazan State Tech University Press, Kazan, 2007) [in Russian].
7. G. S. Settles, *Schlieren and Shadograph Techniques: Visualizing Phenomena in Transparent Media* (Springer, Berlin, 2001).
8. N. N. Evtikhiev, O. A. Evtikhieva, I. N. Kompanets et al., *Information Optics*, Ed. by N. N. Evtikhiev (Moscow Power Engineering Institute Press, Moscow, 2000) [in Russian].
9. J. W. Goodman, *Introduction to Fourier Optics* (McGraw-Hill, New York, 1968; Mir, Moscow, 1970).
10. L. M. Soroko, *Hilbert Optics* (Nauka, Moscow, 1981) [in Russian].
11. V. A. Arbuzov and Y. N. Dubnishchev, *Hilbert Optics Techniques in Measurement Technologies* (Novosibirsk State Tech University Press, Novosibirsk, 2007) [in Russian].
12. M. V. Doroscko, P. P. Kharatsov, O. G. Penyazkov, and I. A. Shikh, "Measurement of Admixture Concentration Fluctuation in a Turbulent Shear Flow Using an Averaged Talbot-Image," in CD-ROM *Proceedings of the 12th International Symposium on Flow Visualization, Goettingen, Sept. 10–14 2006.*
13. G. E. A. Meier, "Computerized Background-Oriented Schlieren," *Experiments in Fluids*, **33**, 181–187 (2002).

14. N. M. Skornyakova, E. M. Popova, B. S. Rinkevichyus, and A. V. Tolkachev, "Correlation Processing of BOS Pictures," in CD-ROM *Proceedings of the 5th International Symposium on Particle Image Velocimetry, Pusan, Korea, 2003*. Paper 3209, p. 11.

15. E. M. Popova, A. V. Tolkachev, and N. M. Skornyakova, "Use of the Background-Oriented Schlieren Method For Natural Convection Studies," in *Optical Methods for Studying Flows: Proceedings of 7th International Conference*. Ed. by Y. N. Dubnishchev and B. S. Rinkevichyus (Moscow Power Engineering Institute Press, Moscow, 2003), pp. 126–129 [in Russian].

16. V. A. Soifer, Ed., *Computer Image Processing Techniques* (Fizmatlit, Moscow, 2001) [in Russian].

17. F. Klinge and J. Kompenhans, "Recent Development and Application of Background Oriented Schlieren Method," in *Optical Methods for Studying Flows: Proceedings of 9th International Conference*. Ed. by Y. N. Dubnishchev and B. S. Rinkevichyus (Moscow Power Engineering Institute Press, Moscow, 2007), pp. 22–25 [in Russian].

18. E. V. Gumennik, *Candidate's Dissertation* (All-Union Scientific Research Institute of Optical-Physics Measurements, Moscow, 1982).

19. E. V. Gumennik and B. S. Rinkevichyus, "The Use of the Refraction of a Scanning Laser Beam for Investigations into the Structure of Transparent Inhomogeneities," *Teplofizika Vysokikh Temperatur*, **25**(6), 1191–1200 (1987).

20. E. V. Gumennik and B. S. Rinkevichyus, "Scanning Laser Refractometer," *Pribory i Tekhnika Eksperimenta*, No. 1, 244 (1990).

21. O. A. Evtikhieva and B. S. Rinkevichyus, RF Patent No. 704339 (4 June 1978).

22. O. A. Evtikhieva, *Candidate's Dissertation* (Moscow Institute of Engineers of Geodezy, Air Surveying and Cartography, Moscow, 1980).

23. N. V. Karlov, *Lectures on Quantum Electronics* (Nauka, Moscow, 1987) [in Russian].

24. M. Francon, *Laser Speckle and Applications in Optics* (Academic Press, New York, 1979; Nauka, Moscow, 1980).

25. N. A. Shatokhina, B. S. Rinkevichyus, and V. A. Zubov, "Using Speckle Interferometry for the Analysis of the Refractive Index Gradient Distribution in Fluid and Gas Flows," in *Laser Anemometry. Advances and Applications*, Ed. by B. Ruck, A. Leder, and D. Dopheide (Karlsruhe, 1997), pp. 399–406.

26. B. S. Rinkevichyus, "Present-Day Laser-Computer Techniques for Gas Flow Diagnostics," in *Fundamental Problems of High-Velocity Flows* (Central Aerohydrodynamics Institute Press, Moscow, 2004), pp. 412–414 [in Russian].

27. M. V. Yesin, O. A. Evtikhieva, S. V. Orlov, B. S. Rinkevichyus, and A. V. Tolkachev, "Laser Refractometral Method for Visualization of Liquid Mixing in Twisted Flows," in CD-ROM *Proceedings of the 10th International Symposium on Flow Visualization, Kyoto, Japan, Aug. 26–29, 2002*. Paper No. F037, pp. 1–8.

28. O. A. Evtikhieva, M. V. Yesin. S. V. Orlov, B. S. Rinkevichyus, and A. V. Tolkachev, "Laser Refraction Method for Studying Liquids in Twisted Flows," in *Proceedings of the 3rd National Conference on Heat Exchange* (Moscow Power Engineering Institute Press, Moscow, 2002), Vol. 1, pp. 197–200 [in Russian].

29. B. S. Rinkevichyus, I. L. Raskovskaya, and A. V. Tolkachev, "Laser Refractography—the New Technology of the Transparent Heterogeneities Quantitative Visualization," in CD-ROM *Proceedings of ISFV13—13th International Symposium on Flow Visualization, FLUVISU12—12th French Congress on Visualization in Fluid Mechanics, Nice, France, July 1–4, 2008*. Paper No. 085.

Chapter 2
Structured Laser Radiation (SLR)

2.1 Main Types of SLR

Structured Laser Radiation is spatially amplitude-modulated radiation obtained with the aid of classical optical elements, DOE, or structured screens [1, 2].

The main elements of structured laser radiation are listed in Table 2.1. These are classified by the shape of the spatial geometrical figures formed by the beams from the source as line-, plane-, or cone-structured laser radiation. The two-dimensional figures presented in the table are cross-sections of the beams formed by families of geometrical optics beams from the source.

By combining the main SLR elements, one can obtain other types of SLR source adapted to the structure of the inhomogeneity at hand and the shape of the body in whose vicinity boundary layers are being studied [3]. To diagnose bulk inhomogeneities, it is advisable to produce "measuring grids" from elementary sources.

Obviously, the idealized representation of SLR cross-sections in the form of geometrical figures is only valid in the geometrical optics approximation, and so when handling actual measuring setups, one should evaluate the errors due to diffraction effects in order that the applicability limits of the method can be determined. For example, plane-structured laser radiation, also referred to as the laser plane, is in fact an astigmatic laser beam of elliptical cross-section whose diffraction divergence is determined by the well-known quasioptical methods [4]. Considered in the following sections are the physical principles of forming simple SLR on the basis of Gaussian laser beams.

B. S. Rinkevichyus et al. (eds.), *Laser Refractography,*
DOI 10.1007/978-1-4419-7397-9_2, © Springer Science+Business Media, LLC 2010

Table 2.1 Main elements of structured laser radiation

Source type	Linear	Plane	Conical
SLR cross-section	·············· ·············· ·············· ··············	▬▬▬▬▬	◯

2.2 Gaussian Beams

2.2.1 Properties of Laser Radiation

Laser radiation differs from radiation produced by the ordinary thermal and luminescent light sources by high coherence, i.e., monochromaticity and directivity. Both these properties are of great importance in laser refractography. The monochromaticity of radiation allows one to disregard the dispersion of the medium, and its narrow directivity makes it possible to produce a new type of radiation, namely, structured radiation, by means of simple optical systems.

The laser is a system consisting of an active (amplifying) medium and a resonator (cavity) comprising one or more high-reflectivity mirrors. If the amplification (gain) of the medium exceeds losses, and the cavity provides for a positive feedback, a narrowly directed monochromatic radiation—the laser beam—is then formed at the exit from the laser [5]. In most cases, this radiation is polarized.

The spatial characteristics of laser radiation are determined by its mode composition. In the laser cavity transverse electromagnetic waves are generated, designated as TEM_{mn} modes, where m and n are integers: $m, n = 1, 2, 3,\dots$. Two lasing (laser generation) regimes are distinguished, namely, multi- and single-mode ones. Single-mode lasing is characterized by sets of indices m and n of one and the same value, whereas multimode generation by those of different values. For a TEM_{mn} mode and rectangular cavity mirrors, the distribution of the electric field amplitude in the beam cross-section is described by the following expression [5]:

$$A(x,y) = A_0 H_m\left(\frac{\sqrt{2}x}{w}\right) H_n\left(\frac{\sqrt{2}y}{w}\right) \exp\left\{-\frac{x^2+y^2}{w^2}\right\}, \qquad (2.1)$$

where A_0 is the coefficient determining the field amplitude, $H_m(x)$ and $H_n(x)$ are Hermitian polynomials of degrees m and n, and w is the laser beam radius. The Hermitian polynomials of the first three degrees have the form

$$H_0(x) = 1, \quad H_1(x) = 2x, \quad H_2(x) = 4x^2 - 2, \quad H_3(x) = 8x^3 - 12x.$$

The field amplitude distribution over the beam cross-section for the first four low-order modes is illustrated in Fig. 2.1.

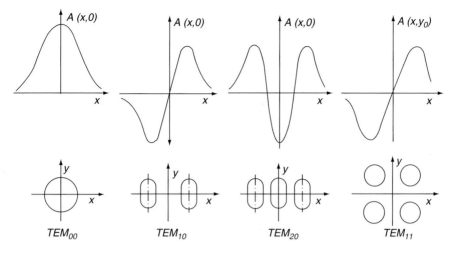

Fig. 2.1 Field amplitude distribution over the beam cross-section for the first four modes

The laser beam radius in expression (2.1) is a scale parameter that depends on the laser cavity configuration and varies along the beam axis. The beam radii w at the mirrors and w_0 at the beam waist, where the beam size is a minimum, are governed by the radii of curvature of the mirrors and the distance between them. If one of the mirrors is flat, the beam waist is located at it; if the mirrors are of the same radius of curvature, the waist is at the center of the cavity.

The laser radiation characteristics considered above are idealized. Actually, deviations are observed from the ideal field amplitude distribution over the beam cross-section and radiation spectrum, due to various kinds of fluctuations in the active medium and cavity parameters. All fluctuations in lasers are customarily classified under two types, natural and technical. Natural fluctuations are due to the atomic structure of the active medium and the quantum character of laser radiation; and technical ones are caused by the slow variations of the cavity and active medium parameters (amplification coefficient and refractive index fluctuations over the cross-section of the laser beam). Technical fluctuations can be reduced by improving the stability of the cavity and the parameters of the active medium.

The spatial coherence of laser radiation means correlation between perturbations at two space-apart points at one and the same instant of time. It affects the visibility of interference patterns and is measured by the Junge method [4]. The complex-degree spatial coherence module for single-mode gas lasers is close to unity. However, account should be taken of the fact that the spatial coherence of laser beams is substantially affected by the medium they pass through.

The divergence of laser beams depends on the shape and arrangement of the cavity mirrors. If these are flat, the divergence is determined by the diffraction loss. The beam divergence in industrial gas lasers ranges between 5 and 20 min, and it can be as great as a few tens of degrees in semiconductor lasers.

The polarization of radiation in gas lasers built around gas-discharge tubes equipped with Brewster-angle windows is linear. The plane of polarization of laser radiation here lies in the plane of incidence of the beam upon the window.

The polarization of the laser beam in gas lasers using gas-discharge tubes with intracavity mirrors depends on the cavity adjustment. As a rule, such lasers use axial magnetic field, which makes possible their two-frequency operation. Radiation of frequencies v_1 and v_2 is circularly polarized.

2.2.2 Characteristics of the Gaussian Beam

The processes of laser diagnostics of flows can most conveniently be analyzed using as an example laser beams of the principal, TEM_{00}, mode that are referred to as Gaussian beams, for the variation of the field amplitude in any cross-section of such a beam is described by a Gaussian curve (Fig. 2.2).

The Gaussian beam is characterized by the following parameters: the beam radius w, the radius of the wave front curvature, $R(z)$, the position of the beam waist where the beam radius is a minimum, the beam waist radius w_0, the beam confocal parameter given by

$$R_0 = kw_0^2/2, \tag{2.2}$$

where k is the wave number, and the far-field divergence angle θ.

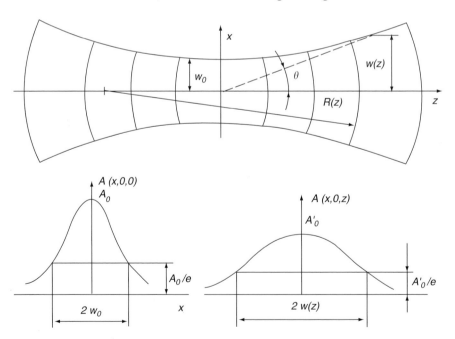

Fig. 2.2 Variation of the parameters of the Gaussian beam along its propagation axis

The properties of Gaussian beams were examined by Evtikhiev and coworkers [4], Karlov [5], and Goncharenko [6]. The radius of such a beam is defined as the distance from the beam axis at which the field amplitude is reduced by a factor of e as compared with the amplitude on the axis. (The beam radius is sometimes defined as the distance at which the field strength is reduced by a factor of e in comparison with the field strength on the beam axis.)

The Gaussian beam has its minimal diameter of $2w_0$ at its waist, where the wave front is flat, or plane. The variation of the beam radius with distance from the waist is described by the following relation:

$$w(z) = w_0 \left[1 + (z/R_0)^2\right]^{1/2}, \qquad (2.3)$$

where z is the distance from the waist. At great distances from the waist, where $(z/R_0)^2 \gg 1$, the beam radius varies with the distance z as

$$w(z) = w_0 z / R_0. \qquad (2.4)$$

The radius $R(z)$ of the wave front curvature depends on the distance z reckoned from the waist:

$$R(z) = z\left[1 + (R_0/z)^2\right]. \qquad (2.5)$$

As the distance from the waist increases, the radius of curvature of the wave front diminishes to assume its minimal value of $R_{min} = 2R_0$ at a distance of $z = R_0$. Thereafter the radius of curvature increases and asymptotically tends to z. Thus, at a great distance from the waist, the wave front of the Gaussian beam takes the form of a sphere with its center at the waist and its radius of curvature equal to z. The region where the radius of curvature decreases with the increasing distance z is called the near field of the Gaussian beam; and the region where this radius increases with the distance z is referred to as the far field of the beam. The near field covers the region $z \leq R_0$ and the far field ranges from $z = R_0$ to infinity. Figure 2.3 shows the relationship between R_0 and w_0 for various lasers. At a great distance from the waist (far field), the Gaussian beam can be characterized by the divergence angle $\theta = w_0 / R_0$.

The distribution of the field strength in the Gaussian beam has the form

$$E(x, y, z, t) = A_0 \left[w_0 / w(z)\right] \exp\{-j\omega t\}$$
$$\times \exp\{j\left[kz + k(x^2 + y^2)/(2R(z)) + \varphi\right] - (x^2 + y^2)/w^2(z)\}. \qquad (2.6)$$

Here A_0 is the field amplitude on the beam axis at the waist, ω is the circular frequency, and $\varphi = \arctan(z/R_0)$ is the phase shift on the beam axis.

The dependences of $w(z)$, φ, and $R(z)$ on the distance z/R_0 are presented in Fig. 2.4.

Fig. 2.3 Confocal parameter as a function of the waist radius of a Gaussian beam: 1—$\lambda=0.488$ μm; 2—$\lambda=0.633$ μm; 3—$\lambda=1.06$ μm

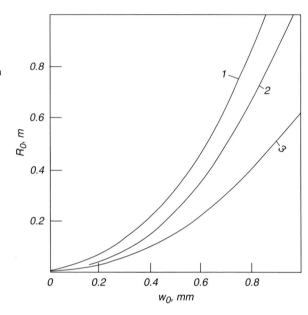

Fig. 2.4 Variation of the Gaussian beam parameters along the z-axis

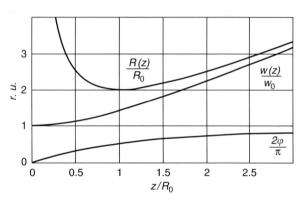

The field amplitude A_0 on the beam axis at the waist may be expressed in terms of the laser beam power P as follows. The distribution of the field strength at the Gaussian beam waist is given by

$$E(x,y,0,t) = A_0 \exp\{-j\omega t\} \exp\{-(x^2+y^2)/w_0^2\},$$

then, according to [2], $P=QA_0^2\pi w_0^2/2$, whence we have

$$A_0 = \sqrt{2P/(\pi w_0^2 Q)}, \qquad (2.7)$$

where $Q=1.35\times10^{-3}$. For instance, if a laser beam 10 mW in power is focused into a spot 50 μm in radius, then $A_0=4.4\times10^4$ V/m.

2.2.3 Propagation of the Gaussian Beam through Optical Elements

Consider the change the Gaussian beam undergoes during the course of its passage through a lens with a focal length of f placed at a distance of l_1 from its waist (Fig. 2.5). Such a lens transforms a Gaussian beam with a waist radius of w_{01} into a beam having a waist radius of w_{02}, this being given by [6]

$$w_{02} = w_{01}\left[\left(1 - l_1/f\right)^2 + R_{01}^2/f^2\right]^{-1/2}. \tag{2.8}$$

The location of the beam waist is defined by the relation

$$1 - l_2/f = \left(1 - l_1/f\right)\left[\left(1 - l_1/f\right)^2 + R_{01}^2/f^2\right]^{-1}. \tag{2.9}$$

In these formulas, the ratios l_1/f and l_2/f are positive if the lens is a collective type and negative if it is dispersive. If the distance l_2 proves negative in calculation, the beam continues to diverge upon leaving the lens. If we put $R_{01}=0$ in expression (2.9), we get the well-known lens formula for homocentric beams.

Analyzing the results obtained, note that the distance l_2 for the Gaussian beam can be shorter or longer than the distance l_2' for the homocentric one, depending on the ratio l_1/f: at $l_1/f>1$ we have $l_2>l_2'$, and if $l_1/f<1$, $l_2<l_2'$.

Let us determine the conditions wherein a lens with a focal length of f transforms the Gaussian beam so that $w_{02}=w_{01}$. It is well known from geometrical optics that the linear magnification of an optical system equals unity if $l_1=2f$. For the Gaussian beam, we have from expression (2.9) that in this case the lens should be moved away from the beam waist for a distance of l_1 such that

$$l_1 = f \pm \left(f^2 - R_{01}^2\right)^{1/2},$$

that is, the distance l_1 depends on the confocal parameter R_{01} of the beam. Specifically, if $R_{01}=f$, the lens should be placed at a distance of f from the beam waist. It is well known that for a homocentric beam in the case where $l_1=f$, a parallel beam is

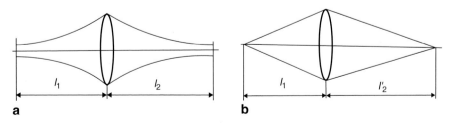

Fig. 2.5 Transforms of a Gaussian and homocentric beams by thin lens. **a** Gaussian beam. **b** homocentric beam

formed at the exit from the lens. At $l_1 = f$, in the case of the Gaussian beam, we have $w_{02} = w_{01} f/R_{01}$; i.e., the magnification of the optical system is governed by the ratio between the focal length of the lens and the confocal parameter of the beam. The beam waist radius after the lens can be either smaller or greater than that before the lens. It follows from expression (2.8) that to minimize the spot size, one should take a short-focus lens and move it away for a great distance. If $l_1 \gg R_{01} - f$, then $w_{02} = w_{01} f/(f + l_1)$. But one should bear in mind here that a beam waist less than 5 μm in size is difficult to obtain because of the aberrations of the lens and the diffraction phenomenon.

It frequently proves necessary to transform a Gaussian beam with a waist radius of w_{01} into a beam with a waist radius of w_{02} by means of a lens with a focal length of f. Such transformation takes place if the lens is moved away from the waist for a distance of l_1 such that

$$l_1 = f \pm R_{01}\left(f^2/f_0^2 - 1\right)^{1/2}, \tag{2.10}$$

the waist with a radius of w_{02} occurring at a distance of l_2 from the lens, given by

$$l_2 = f \pm R_{02}\left(f^2/f_0^2 - 1\right)^{1/2}, \tag{2.11}$$

where $f_0^2 = R_{01} R_{02}$, and the signs in the formulas are the same. It should be noted here that such a transformation is only possible if $f \geq f_0$.

More often than not, it is the inverse problem that is being solved in practice, namely, which lens one should take and how to place it in order to obtain a beam waist of desired size at a specified distance. From formulas (2.8) and (2.9) we get the following expressions to calculate the focal length of the lens and its location, given the original waist radius w_{01}, the waist radius w_{02} that is necessary to obtain, and the distance l_2 between the lens and the waist:

$$f = \left[l_2 R_{01} - \sqrt{R_{01} R_{02}\left(R_{02}^2 - R_{01} R_{02} + l_2^2\right)}\right] \Big/ (R_{01} - R_{02}),$$

$$l_1 = f + \left[R_{01}\left(f^2 - R_{01} R_{02}\right)/R_{02}\right]^{1/2}.$$

A single lens can help achieve the parameters specified if

$$l_2^2 \geq R_{02}^2\left[R_{01}/R_{02} - 1\right].$$

This means that it is not always possible to obtain a Gaussian beam with a waist of desired size by means of a single lens.

The double-lens system shown schematically in Fig. 2.6 has great functional capabilities.

The transformation of the Gaussian beam in this system is described by the following expressions [7]:

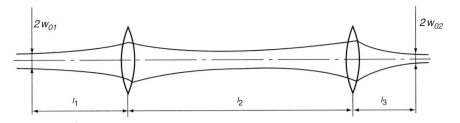

Fig. 2.6 Transformation of the Gaussian beam with a double-lens system

$$R_{03} = f_1^2 f_2^2 R_{01} / \left\{ \left[(f_1 - l_1)(l_2 - f_2) + l_1 f_2 \right]^2 + R_{01}^2 (f_1 + f_2 - l_2)^2 \right\},$$

$$l_3 = \frac{f_2 \left\{ [f_1 l_1 + l_2 (f_1 - l_1)] \left[(f_1 - l_1)(l_2 - f_2) + l_1 f_2 \right] + R_{01}^2 (f_1 - l_2)(f_1 + f_2 - l_2) \right\}}{\left[(f_1 - l_1)(l_2 - f_2) + l_1 f_1 \right]^2 + R_{01}^2 (f_1 + f_2 - l_2)^2},$$

where f_1 and f_2 are the focal lengths of the first and the second lens, respectively, l_1 is the distance from the waist of the original Gaussian beam to the first lens, l_2 is the distance from the first to the second lens, and l_3 is the distance from the second lens to the new beam waist.

These formulas can also help obtain the relations necessary for the solution of the inverse problem, namely, what are the lens parameters f_1 and f_2 and the distances l_1, l_2, and l_3 necessary to obtain the desired waist radius w_{03}, given the waist radius w_{01}. Calculation examples for a simple double-lens optical system capable of ensuring Gaussian beam parameters, unattainable with single-lens systems can be found in [7].

When calculating the characteristics of optical schemes for laser refractographic systems, one has to determine the parameters of the Gaussian beam passing through inclined plane-parallel plates, for example, windows, which limit the flow of interest, and individual regions of the medium under study that differ in refractive index, whose interface is inclined relative to the incident beam axis. In this case, the Gaussian beam becomes astigmatic; i.e., the waist positions of the beam and its radii of curvature in the meridional and sagittal planes fail to coincide. The formulas to calculate the beam parameters in such cases can be found in [7]. The classical Gaussian beams serve as a basis for formation, with the help of optical elements, the simplest types of SLR that are considered in the subsequent sections of the book.

2.3 Formation of SLR on the Basis of Optical Elements

2.3.1 Formation of the Laser Plane

Laser refractographs based on astigmatic laser beams—laser planes—are at present the most widespread type. Depending on the problem at hand, the requirements for the parameters of the laser plane differ widely, and so various systems are being used to form it.

Single-Lens System. Consider the characteristics of the single-lens system used to form the laser plane (Fig. 2.7). Placed in the path of the low-divergence beam of laser *1* is a negative or positive cylindrical short-focus lens *2* that first focuses the beam and then widens it in a single plane only.

For an aberration-free lens, the width $h(z)$ of the laser plane along the x-axis and its thickness $t(z)$ along the y-axis are calculated by the following formulas:

$$h(L) = 2w_1\left[1 + \left(\frac{l_2 - l_1}{R_{01}}\right)^2\right]^{1/2}, \tag{2.12}$$

$$t(L) = 2w_0\left[1 + \left(\frac{L}{R_0}\right)^2\right]^{1/2}, \tag{2.13}$$

where w_0 is the waist radius of the original laser beam, w_1 is the waist radius of the beam after the lens, l_0 is the distance from the waist of the original laser beam to the lens, l_1 is the distance from the lens to the waist of the beam after the lens, l_2 is the distance from the lens to the image recording plane, R_0 is the confocal parameter of the original laser beam, and R_{01} is the confocal parameter of the focused laser beam. The confocal parameters R_0 and R_{01} are given by

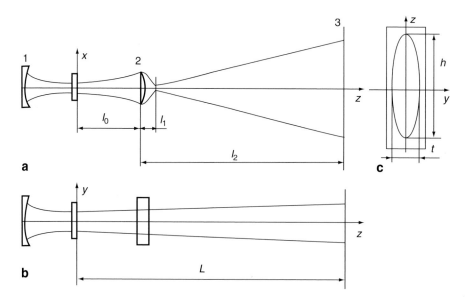

Fig. 2.7 Single-lens optical system for forming the laser plane. **a** View in the *xoz* plane. **b** View in the *yoz* plane. **c** Cross-section of the SLR formed. *1*—laser; *2*—cylindrical lens; *3*—image recording plane

$$R_0 = \frac{\pi w_0^2}{\lambda}, \quad R_{01} = \frac{\pi w_1^2}{\lambda}.$$

The radius w_1 of the new waist is calculated by the Gaussian beam transformation formula 2

$$w_1 = w_0 \left[\left(1 - \frac{l_0}{f}\right)^2 + \frac{R_0^2}{f^2} \right]^{-1/2},$$

where f is the focal length of the cylindrical lens.

The position of the waist of the beam after the lens is defined by the relation:

$$1 - \frac{l_1}{f} = \left(1 - \frac{l_0}{f}\right) \left[\left(1 - \frac{l_0}{f}\right)^2 + \frac{R_0^2}{f^2} \right]^{-1}. \tag{2.14}$$

If $R_{01} > R_0$, the divergence of the astigmatic beam after the lens along the x-axis will be greater than that along the y-axis. The ratio between the dimensions of this beam is

$$\frac{h(L)}{t(L)} = \frac{\left[1 + \left(\dfrac{L - l_0 - l_1}{R_{01}}\right)^2 \right]^{1/2}}{\left[1 + \left(\dfrac{L}{R_0}\right)^2 \right]^{1/2} \left[\left(1 - \dfrac{l_0}{f}\right)^2 + \left(\dfrac{R_0}{f}\right)^2 \right]^{1/2}}. \tag{2.15}$$

Figure 2.8 shows the astigmatism of a Gaussian beam as a function of its propagation distance.

The single-lens laser plane forming system has the advantage of simplicity, but suffers from a substantial shortcoming: because of the strong divergence of the beam, its power density lowers materially as its propagation distance grows longer.

Double-Lens System. A laser plane with a small divergence angle can be obtained by means of two cylindrical lenses forming a telescopic system (Fig. 2.9).

The dimensions of an astigmatic beam are defined by the relations

$$h(L) = 2w_1 \left[1 + \left(\frac{l_2 - l_1}{R_{01}}\right)^2 \right]^{1/2}, \quad t(L) = 2w_0 \left[1 + \left(\frac{L}{R_0}\right)^2 \right]^{1/2},$$

where $L = l_0 + (n-1)h_1 + l_1 + l_2 + (n-1)h_2 + l_3$, h_1 is the thickness of the first lens, h_2 is the thickness of the second lens, and n is the refractive index of the lenses. The distances l_1 and l_2 are calculated by the Gaussian beam transformation formulas.

Triple-Lens Optical System. In the single- and double-lens optical systems analyzed above, the beam thickness in the xoz plane at a distance of L is governed by

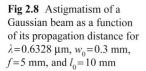

Fig 2.8 Astigmatism of a Gaussian beam as a function of its propagation distance for $\lambda=0.6328$ μm, $w_0=0.3$ mm, $f=5$ mm, and $l_0=10$ mm

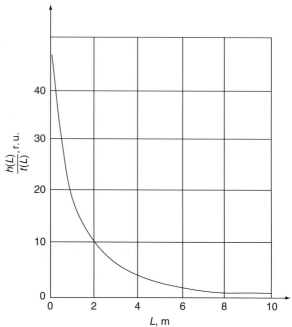

the divergence of the beam issuing from the laser. To reduce the thickness of the laser plane at the location of the optical inhomogeneity of interest, it is necessary to additionally use a spherical long-focus lens. The optical scheme of the triple-lens system for forming the laser plane is presented in Fig. 2.10. It comprises three

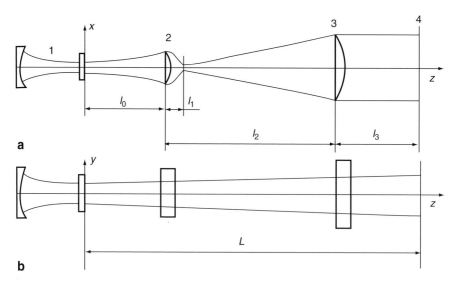

Fig. 2.9 Optical system to form a low-divergence laser plane. **a** View in the *xoz* plane. **b** View in the *yoz* plane. *1*—laser; *2* and *3*—cylindrical lenses; *4*—plane of observation

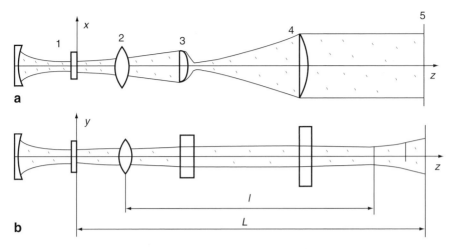

Fig. 2.10 Triple-lens optical system for forming the laser plane. **a** View in the *xoz* plane. **b** View in the *yoz* plane. *1*—laser; *2*—spherical lens; *3* and *4*—cylindrical lenses; *5*—plane of observation

lenses: a spherical lens forming a beam waist at a great distance *l* and two cylindrical lenses that form the laser plane proper.

2.3.2 Besselian Beams and Their Formation

The Besselian beam is a laser radiation model having a number of specific features [8], namely, the distribution of intensity over its cross-section is described with the aid of Bessel functions of varying order, the beam is centrally symmetric and retains its cross-sectional structure during the course of propagation (in practice, this property maintains only for a certain distance from the radiation source). Such beams were first studied and obtained experimentally in 1987.

Distinguished are several kinds of Besselian beams, from the simplest classical Besselian beam described by a Bessel function of order zero to complex tubular models that are classed with Besselian beams only as regards the complete set of the attributes described above [9].

The use of Besselian beams in the diagnostics of optically inhomogeneous media holds much promise, thanks to their high spatial properties (high spatial concentration of radiation within a narrow laser pencil or tubular beam). This makes it possible to obtain narrow laser probes, and study with them inhomogeneities in media at long distances.

There are several optical schemes for forming Besselian beams, most widespread among them being those built around conical lenses—axicons—and round apertures [9].

The field of a Besselian beam propagating along the *z*-axis in the cylindrical coordinate system (r, ψ, z) is described as follows:

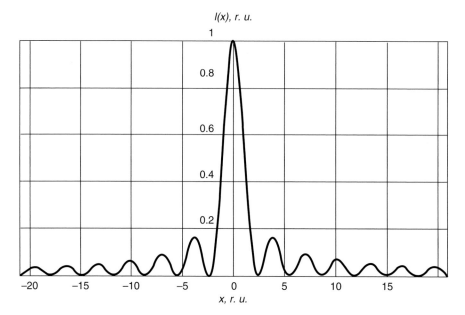

Fig. 2.11 Transverse intensity distribution in the Besselian beam

$$E(r, \psi, z) = A J_0 \left(k_0 n r \sin \theta_0 \right) \exp \left(j k_0 n z \cos \theta_0 \right), \tag{2.16}$$

where A is the maximum field amplitude, $J_0(x)$ is the Bessel function of the first kind of order zero, θ_0 is the angular parameter of the beam, n is the refractive index of the medium wherein the beam propagates, and k_0 is the wave number in vacuum. The beam amplitude is independent of the angle ψ; i.e., it is centrally symmetric. Figure 2.11 shows the transverse intensity distribution in the zero-order Besselian beam.

The Besselian beam is peculiar in that its phase varies in the z-direction at a rate of $v = c/(n \cos \theta_0)$ higher than $v = c/n$, the shape of the beam remaining unchanged during propagation (the amplitude being independent of z). This situation is akin to the one wherein the interference of two plane waves is observed, the only difference being that the interference field fails to occupy the entire space, but is practically wholly concentrated in a certain zone ($k_0 n r \sin \theta_0 << 2.4$, where 2.4 is the first zero of the function J_0). Thus, the Besselian beam can be represented as the result of interference of a multitude of plane waves propagating at an angle of θ_0 to the z-axis. Let us demonstrate how expression (2.16) can be derived from this assumption.

Let us assume that in the spherical coordinate system ρ, θ, φ, where the angle $\theta = 0$ corresponds to the positive direction of the z-axis, there are a set of plane waves of the same amplitude $A d\varphi$, whose propagation directions, defined as $s = \alpha i + \beta j + \gamma k$, make the same angle θ_0 with the z-axis. Each of such fields may be written down in the rectangular coordinate system as

$$E(x, y, z) = A \exp(jk_0 n(\alpha x + \beta y + yz)) \, d\varphi, \qquad (2.17)$$

where $\alpha = \sin\theta_0 \cos\varphi$, $\beta = \sin\theta_0 \sin\varphi$, $\gamma = \cos\theta_0$. If we integrate expression (2.17) with respect to $d\varphi$ from 0 to 2π, then, with the properties of the Bessel function (the central symmetry of the beam and the presence of the principal maximum at the center) known, we arrive at expression (2.16):

$$A \int_0^{2\pi} \exp(jk_0 n(\alpha x + \beta y + \gamma z)) \, d\varphi$$

$$= A \exp(jk_0 nz \cos\theta_0) \int_0^{2\pi} A \exp(jk_0 n\rho(\cos\varphi \cos\psi + \sin\varphi \sin\psi) \sin\theta_0) \, d\varphi$$

$$= 2\pi A J_0(k_0 n\rho \sin\theta_0) \exp(jk_0 nz \cos\theta_0). \qquad (2.18)$$

One of the most important problems associated with Besselian beams is the problem of their formation. A most simple way to obtain a Besselian beam is to use an axicon (Fig. 2.12). If a conical lens—an axicon—is placed in the path of a wide Gaussian beam, the beams deflected by it will interfere, producing a complex intensity distribution pattern. It will have the form of a bright central maximum surrounded by a system of rings.

In numerical modeling, one can use Besselian beam models specified in the MathCAD program by Bessel functions of order n, $J_n\left(k \sin(\theta) \sqrt{x^2 + y^2}\right)$. In the MathCAD environment, the function $J_n(x)$ is sought as the solution of the differential equation $x^2(d^2/dx^2)y + x(d/dx)y + (x^2 - n^2)y = 0$, where the function $y(x)$ is the solution. In addition, the beam is specified by the wave number k and the axicon angle θ. The axicon angle is the parameter determining the cross-sectional size of the beam. The physical meaning of this angle is illustrated in Fig. 2.13. In the general case, the Besselian beam results from the interference of plane waves propagating, thanks to the axicon, at the same angle to the optical axis, their propagation directions forming a cone. Its apex angle is the axicon

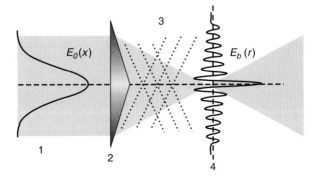

Fig. 2.12 Principle of formation of the Besselian beam with the axicon: *1*—amplitude distribution in the Gaussian beam; *2*—axicon; *3*—interference field behind the axicon; *4*—amplitude distribution in the Besselian beam

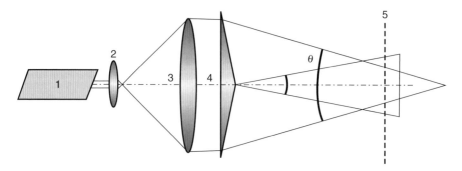

Fig. 2.13 Optical scheme of formation of the Besselian beam: *1*—single-mode laser; *2* and *3*—lenses; *4*—axicon; *5*—beam interference region

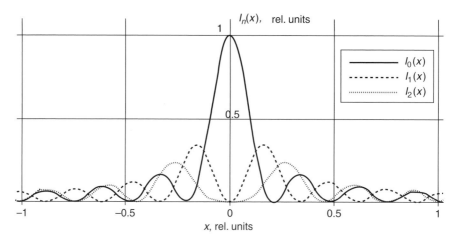

Fig. 2.14 Intensity distribution in the cross-section of the zero-, first-, and second-order Besselian beams

angle θ. Figure 2.14 shows the intensity distribution for Besselian beams of the first three orders.

One can see from the above plots that the principal Besselian mode—the zero-order beam—has its principal maximum at the center, whereas the modes of higher orders are tubular beams, the beam "smearing out" with increasing mode order. Note that as the axicon angle increases, the cross-section of the beams narrows.

2.4 Formation of SLR on the Basis of Diffraction Gratings

Amplitude Gratings. Multiple-beam SLR can be obtained by means of diffraction gratings and an objective lens. Use can be made of both one- and two-dimensional amplitude and phase gratings. Figure 2.15a shows the optical scheme of formation

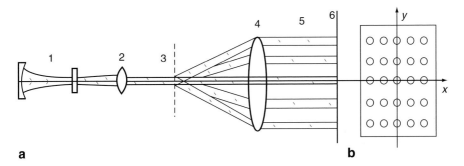

Fig. 2.15 Formation of SLR by means of an amplitude diffraction grating. **a** Optical system. **b** Cross-section of the SLR formed. *1*—laser; *2*—long-focus lens; *3*—two-dimensional diffraction grating; *4*—fast lens; *5*—laser beams; *6*—plane of observation

of a multitude of individual beams with a two-dimensional amplitude grating. Here radiation from laser *1* successively passes through long-focus collective lens *2*, diffraction grating *3*, and fast lens *4* at whose exit a set of parallel laser beams *5* are formed that can be observed on diffusing screen *6*. The cross-sectional appearance of the SLR formed is shown in Fig. 2.15b.

As demonstrated in Sect. 2.12, the double-lens system can make laser beams have the desired size at the location of the optical inhomogeneity under study in an optimal way.

The distances Λ_x and Λ_y between the beams along the *x*- and *y*-axes, respectively, are governed by the periods of the diffraction grating and the focal length f of the objective lens: $\Lambda_x = \lambda f/d_x$, $\Lambda_y = \lambda f/d_y$, where d_x and d_y are the periods of the grating along the *x*- and *y*-axes, respectively.

Phase Gratings. SLR can be formed by means of dynamic phase gratings. Such gratings are formed owing to the acousto-optic effect occurring during propagation of ultrasonic waves in transparent media. Such media may be represented by liquids, glasses, and crystals. Ultrasonic waves are generated by piezoelectric emitters. The Raman-Nath diffraction observed in the frequency range 3–20 MHz yields a multitude of diffracted beams. Figure 2.16 illustrates the formation of SLR by means of a phase diffraction grating. Here radiation from laser *1* passes through long-focus lens *2* and ultrasonic modulator *3* supplied from high-frequency generator *4*. Following the modulator there a set of diffracted divergent beams appear that are transformed with objective lens *5* into SLR *6* observed on screen *7*.

When ultrasonic waves 10 MHz in frequency propagate in water with a velocity of 1,500 m/s, a periodic variation of the refractive index of the medium takes place, the variation period being equal to the length of the acoustic wave, 150 μm in our case. The angle the first diffracted wave makes with the original beam (the diffraction angle) for a wavelength of $\lambda = 0.6328$ μm amounts to 0.0042 rad. If use is made of an objective lens 500 mm in focal length, the distance between the parallel beams will be 2.1 mm. The number of diffracted waves depends on the power capacity of the generator and can come to a few tens.

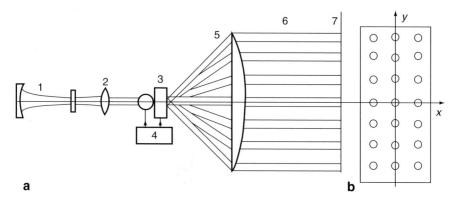

Fig. 2.16 Formation of SLR by means of a phase diffraction grating. **a** Optical system. **b** Cross-section of the SLR formed. *1*—laser; *2*—spherical lens; *3*—two-coordinate ultrasonic modulator; *4*—high-frequency generator; *5*—cylindrical lens; *6*—laser beams; *7*—plane of observation

This method of formation of linear SLR has the advantages of simplicity and capability of readily changing the distance between the parallel beams by varying the frequency of the generator. The shortcomings include the nonuniform distribution of laser power among the diffracted beams and narrow variation range of the grating period.

2.5 Formation of SLR on the Basis of Diffraction Elements

The diffraction optical elements (DOEs) that have only recently become commercially available have the form of a thin phase plate with a special phase relief laser engraved in it [10]. The diffraction of laser radiation by such an optical element produces various kinds of spatially modulated radiation known as structured laser radiation. DOEs are used with both gas and semiconductor lasers generating highly astigmatic beams. DOEs can help obtain SLR of practically any structure, adapted for solving problems in the diagnosis of fields of gradient refractive indices and other physical quantities.

At present, DOEs are used to split laser radiation, in spectroscopy, metrology, and multi-perspective measuring systems. DOEs were first used to obtain SLR for studies into optically inhomogeneous media, and also for their visualization from scattered radiation in [1, 2]. The original elliptical or high-divergence beam from a semiconductor laser is transformed with a corrector lens into an almost axially symmetric beam that is next converted with a diffraction optical element equivalent to a linear diffraction grating into a set of divergent beams arranged in a single plane. The angle between the adjacent beams is $\Theta = \lambda/\Lambda$, the total divergence angle being given by

$$\Theta_{tot} = (2m + 1)(\lambda/\Lambda),$$

where $2m+1$ is the number of diffracted beams, m is the diffraction order, λ is the laser radiation wavelength, and Λ is the period of the diffraction grating. To illustrate, at $m=5$, $\lambda=0.68$ µm, and $\Lambda=6.8$ mm the total divergence angle for eleven beams is $\Theta_{tot}=27°$.

A diffraction optical element equivalent to a cylindrical lens can help obtain from a laser beam a divergent laser plane whose cross-section is a straight line. A cylindrical-lens-cum-diffraction-grating combination makes it possible to obtain a number of such lines, usually 5, 9, 11, 15, 33, 99. The number of lines can exceed 100. These beams are divergent. To illustrate, for a diffraction optical element with a period of 200 µm that produces 99 lines, the angle between the adjacent lines amounts to 0.15°, the total divergence angle being equal to 14.47°.

The use of DOEs with more intricate diffraction reliefs allows obtaining SLR of more complex forms: a cross-shaped beam comprising two laser planes arranged at right angles, a conical beam, or a set of conical beams, etc. [10].

Table 2.2 presents the main types of SLR obtained with commercially available DOEs [3].

Table 2.2 Main types of SLR obtained with DOEs

Line	Cross	Parallel lines
Square	Dotted line	7 by 7 matrix
Ring	7 concentric rings	Dot

References

1. O. A. Evtikhieva, B. S. Rinkevichyus, and A. V. Tolkachev, "Visualization of Nonstationary Convection in Liquids Near Hot Bodies with Structured Laser Beam," in CD-ROM *Proceedings of the 12th International Symposium on Flow Visualization, Goettingen, Sept. 10–14, 2006.*
2. I. L. Raskovskaya, "Structured beams in laser refractography," *Journal of Communications Technology and Electronics* **54**(12), pp. 1445–1452 (2009).
3. http://www.ProPhotonix Limited
4. N. N. Evtikhiev, O. A. Evtikhieva, I. N. Kompanets *et al., Information Optics*, Ed. by N. N. Evtikhiev (Moscow Power Engineering Institute Press, Moscow, 2000) [in Russian].
5. N. V. Karlov, *Lectures on Quantum Electronics* (Nauka, Moscow, 1983) [in Russian].
6. A. M. Gonchaenko, *Gaussian Light Beams* (Nauka i Tekhnika, Minsk, 1977) [in Russian].
7. B. S. Rinkevichius, *Laser Diagnostics in Fluid Mechanics* (Begell House Inc. Publishers, New York, 1998).
8. N. E. Andreev, S. S. Bychkov, V. V. Kotlyar *et al.*, "Formation of High-Power Tubular Besselian Light Beams," *Kvantovaya Elektronika* **23**(2), pp. 130–134 (1996).
9. V. V. Korobkin, L. Y. Polonsky, V. P. Poponin, and L. N. Pyatnitsky, "Focusing of Gaussian and Hyper-Gaussian Laser Beams with Axicons to Obtain Continuous Laser Sparks," *Kvantovaya Elektronika* **13**(2), pp. 265–270 (1986).
10. V. A. Soifer, Ed., *Computer Optics Techniques* (Fizmatlit, Moscow, 2003).

Chapter 3
Physical Processes Giving Rise to Optical Inhomogeneities in Media

3.1 Temperature Field in Liquids

Note that all optical methods allow information to be obtained only about the refractive index field that next can be converted, by the appropriate calculation procedures, into the temperature field $T(x, y)$, or into the field of some other physical quantity of interest. In mixed heat and mass transfer processes, the variation of the refractive index depends on two quantities, as defined by the relation [1]

$$dn = \frac{\partial n}{\partial T}dT + \frac{\partial n}{\partial C}dC, \tag{3.1}$$

where dn/dT is the refractive index derivative with respect to the temperature T and dn/dC is that with respect to the concentration C. Even if both the partial derivatives in Eq. (3.1) are known, it nevertheless proves generally difficult to obtain the temperature or concentration field solely from the refractive index field measured. One frequently tries to find one of the two fields theoretically or by nonoptical measurement methods. Besides, by selecting suitable working medium, one can satisfy the following relation:

$$\frac{\partial n}{\partial T}\Delta T \gg \frac{\partial n}{\partial C}\Delta C, \tag{3.2}$$

where ΔT and ΔC are the greatest differences in temperature and concentration, respectively, say, between the wall and the flow. In the case of nonstationary problems, use can be made of the fact that the two processes take their course at different rates, for the thermal conductivity coefficient in liquids is much higher than the diffusion coefficient.

Table 3.1 lists the refractive indices of some gases and liquids, as well as their temperature coefficients [2]. The refractive index temperature coefficient is proportional to the density temperature coefficient. Since all liquids expand on heating, their refractive indices decrease with increasing temperature.

The refractive index temperature coefficient for most liquids ranges within narrow limits from −0.0004 to −0.0006 1/degree. Important exclusions include water and dilute aqueous solutions (−0.0001), glycerin (−0.0002), and glycol (−0.00026).

B. S. Rinkevichyus et al. (eds.), *Laser Refractography*,
DOI 10.1007/978-1-4419-7397-9_3, © Springer Science+Business Media, LLC 2010

Table 3.1 Refractive indices of some gases and liquids for $\lambda = 0.6328 \ \mu m$

Nos.	Medium	N	$dn/dT \ (°C^{-1})$
1	Air	1.0002724	0.927×10^{-6}
2	Nitrogen	1.0002793	0.949×10^{-6}
3	Oxygen	1.0002531	0.864×10^{-6}
4	Carbon monoxide	1.0004197	1.424×10^{-6}
5	Water vapor	1.0002354	0.798×10^{-6}
6	Water	1.3314	-0.985×10^{-4}
7	Methyl alcohol	1.3253	-4×10^{-4}
8	Benzene	1.495	-6.4×10^{-4}
9	Acetone	1.3542	-5.31×10^{-4}
10	Carbon disulfide	1.6185	-7.96×10^{-4}

Linear extrapolation of the refractive index is permissible to small temperature differences (10–20°C) only. The exact refractive index values within wide temperature limits can be found by empirical formulas of the form

$$n(T) = n_0 + aT + bT^2 + \cdots,$$

where a and b are constants.

The temperature dependence of the refractive index of water for laser radiation with a wavelength of $\lambda = 0.6328 \ \mu m$ is determined by the approximation relation

$$n(T) = 1.3328 - 0.000051 \ T - 0.0000011 \ T^2, \tag{3.3}$$

obtained on the basis of the dispersion formula and the data presented by Rinkevichyus [3] (see Table 3.2).

To select an appropriate model for temperature variation in the vicinity of hot or cold solids in liquids, it seems but natural to use the application packages utilized in

Table 3.2 Refractive indices of distilled water for $\lambda_D = 589.3$ nm

$T \ (°C)$	n_D	$T \ (°C)$	n_D	$T \ (°C)$	n_D	$T \ (°C)$	n_D
0	1.33395	15	1.33339	30	1.33194	45	1.32985
1	1.33395	16	1.33331	31	1.33182	46	1.32969
2	1.33394	17	1.33324	32	1.33170	47	1.32953
3	1.33393	18	1.33316	33	1.33157	48	1.32937
4	1.33391	19	1.33307	34	1.33144	49	1.32920
5	1.33388	20	1.33299	35	1.33131	50	1.32904
6	1.33385	21	1.33290	36	1.33117	51	1.32887
7	1.33382	22	1.33280	37	1.33104	52	1.32870
8	1.33378	23	1.33271	38	1.33090	53	1.32852
9	1.33374	24	1.33261	39	1.33075	54	1.32835
10	1.33369	25	1.33250	40	1.33061	55	1.32817
11	1.33364	26	1.33240	41	1.33046	56	1.32799
12	1.33358	27	1.33229	42	1.33031	57	1.32781
13	1.33352	28	1.33217	43	1.33016	58	1.32762
14	1.33346	29	1.33206	44	1.33001	59	1.32744
–	–	–	–	–	–	60	1.32725

Fig. 3.1 Computer visualization of the temperature field near a hot cylinder in water

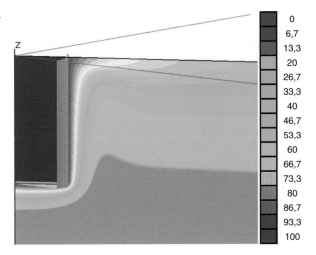

the numerical analysis of thermophysical processes. When calculating temperature fields in liquids with free convection, use is usually made of the Boussinesq approximation [4].

According to this approximation:

1. The density of the liquid is taken to be constant in all the equations describing motion and heat exchange in the medium.
2. A term allowing for the effect of temperature gradient on the motion of the liquid is added to the equations of motion.

Figures 3.1 and 3.2 show the temperature field round a hot flat-bottomed cylinder 34.5 mm in diameter and 72 mm high in water [5], computed with the aid of the ANES–NE program package developed by V. I. Artemov and G. G. Yankov at the Chair of Engineering Thermophysics of Moscow Power Engineering Institute.

Based on the computations performed, one can select the model of the temperature field distribution at the surface of hot or cold solids in liquids. For example, the radial temperature dependence in the spherically symmetric temperature field of a hot ball of radius R is defined by the expression [5]

$$T(r) = T_0 + \Delta T \exp\left(-\frac{(r - R - \Delta R)^2}{a^2}\right), \tag{3.4}$$

where T_0, ΔT, ΔR, and a are the parameters of the temperature field model. The parameter T_0 is determined by the wall temperature of the liquid-filled cell, $T(R)$ is the temperature at the surface of the ball, and the ratio $\Delta T/a$ corresponds to the temperature field gradient in the boundary layer of thickness a. The temperature gradient at $r = R$ is governed by the shift ΔR, and where the heat conductivity of the ball differs from that of the liquid, grad $[T(r)] \neq 0$ at the surface of the ball. However, the possibility cannot be ruled out in investigations that the temperature gradient in the boundary layer will prove to be equal or close to zero, providing reason enough to select the Gaussian temperature field model.

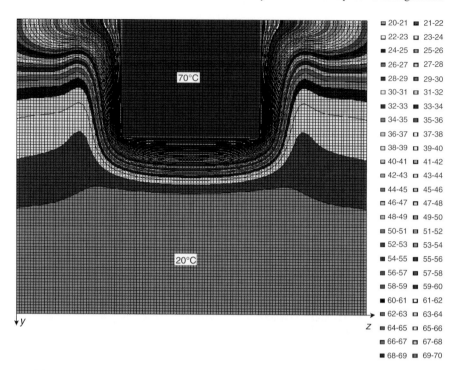

⊡ 20-21	■ 21-22		
▫ 22-23	▫ 23-24		
■ 24-25	⊡ 25-26		
⊡ 26-27	▫ 27-28		
■ 28-29	■ 29-30		
▫ 30-31	▫ 31-32		
■ 32-33	■ 33-34		
■ 34-35	■ 35-36		
⊡ 36-37	▫ 37-38		
▫ 38-39	▫ 39-40		
⊡ 40-41	■ 41-42		
⊡ 42-43	▫ 43-44		
■ 44-45	▫ 45-46		
⊡ 46-47	▫ 47-48		
⊡ 48-49	■ 49-50		
■ 50-51	■ 51-52		
■ 52-53	■ 53-54		
■ 54-55	■ 55-56		
■ 56-57	■ 57-58		
■ 58-59	■ 59-60		
■ 60-61	▫ 61-62		
■ 62-63	▫ 63-64		
■ 64-65	▫ 65-66		
■ 66-67	▫ 67-68		
■ 68-69	▫ 69-70		

Fig. 3.2 Computer visualization of the temperature field near a hot cylinder in water. $T_{cylinder}=70°C$; $T_{water}=20°C$

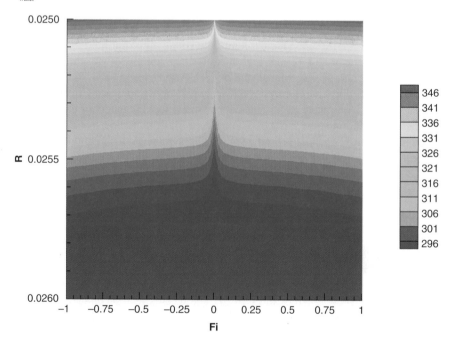

Fig. 3.3 Computer visualization of the boundary layer beneath a hot ball in water: the *vertical scale* is marked off in meters, and the *horizontal one*, in radians

Nevertheless, model (3.4) cannot be used at singular points and regions near the surface of the sphere that are revealed in the vicinity of the vertical symmetry axis through numerical modeling of free convection in the liquid. In Fig. 3.3, such region is visualized in the boundary layer beneath a hot ball in a liquid with the aid of the FLUENT application package. The computation was performed for a metal ball 50 mm in diameter. Plotted on the horizontal axis are the values (in radians) of the angles reckoned from the vertical axis. The computations illustrated by this and subsequent figures (Figs. 3.3–3.6) were made by D. E. Pudovikov.

In actual situations, however, these regions have a more complex, irregular structure, and to study them requires developing special experimental techniques.

Of special interest are investigations into the edge effects occurring in the boundary layers near edges or spikes of hot or cold bodies. Figure 3.4 presents the results of computer visualization of the varying temperature field near an edge of a

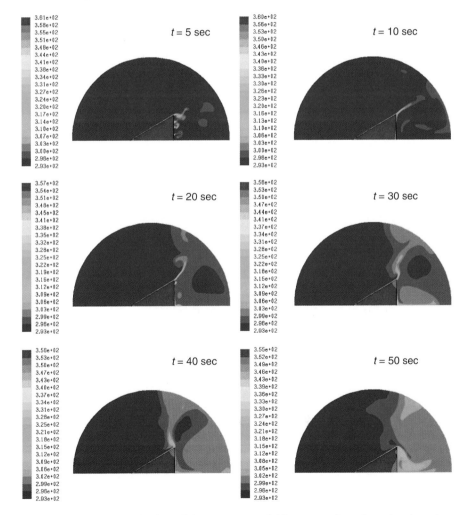

Fig. 3.4 Computer visualization of the temperature field near an edge of a prism heated to $T = 90°C$ at various instants of time

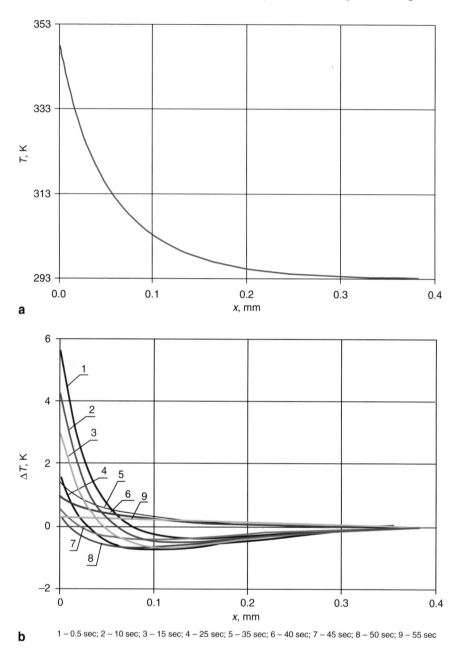

Fig. 3.5 Temperature distribution in the boundary layer near the edge of the prism at various instants of time. **a** At $t = 60$ sec. **b** Deviation of temperature from this distribution at other instants of time

b   1 – 0.5 sec; 2 – 10 sec; 3 – 15 sec; 4 – 25 sec; 5 – 35 sec; 6 – 40 sec; 7 – 45 sec; 8 – 50 sec; 9 – 55 sec

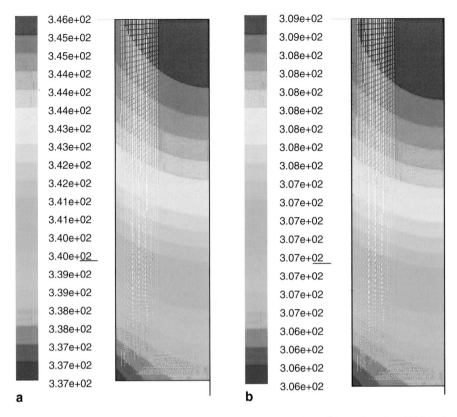

3.46e+02	3.09e+02
3.45e+02	3.09e+02
3.45e+02	3.08e+02
3.44e+02	3.08e+02
3.44e+02	3.08e+02
3.44e+02	3.08e+02
3.43e+02	3.08e+02
3.43e+02	3.08e+02
3.42e+02	3.08e+02
3.42e+02	3.07e+02
3.41e+02	3.07e+02
3.41e+02	3.07e+02
3.40e+02	3.07e+02
3.40e+02	3.07e+02
3.39e+02	3.07e+02
3.39e+02	3.07e+02
3.38e+02	3.07e+02
3.38e+02	3.06e+02
3.37e+02	3.06e+02
3.37e+02	3.06e+02
3.37e+02	3.06e+02
a	**b**

Fig. 3.6 Temperature distribution inside and on the surface of a rod heated to $T = 90°C$ and immersed in a liquid for two instants of time. **a** $t = 10$ sec. **b** $t = 30$ sec

horizontal prism cooling in a liquid for various instants of time, and Fig. 3.5 illustrates the temperature distribution along the symmetry axis in the boundary layer near the edge of the prism.

The nonuniform character of cooling of a hot body in conditions of free convection can be comprehended from Fig. 3.6 illustrating as an example the cooling of a rod, as evidenced by its temperature distribution inside and on the surface.

The computed data are used for preliminary analysis purposes in experimental refractographic studies of actual temperature fields in the situations considered.

3.2 Acoustic Field in Liquids and Gases

Laser methods for the diagnosis of acoustic fields can be used in investigations into natural atmospheric and hydroacoustic fields, in designing hydroacoustic communication channels, in hydrophone calibration, and in other metrological applications. The development of new methods is being stimulated by the research on the effect of ultrasound on matter and studies of untraditional acoustic wave sources and generation mechanisms. The latter include the dynamic, strictional, Cherenkov, bubble, thermal, and microshock wave generation mechanisms.

The contactless laser methods are an indispensable tool in investigations into the sonoluminescence and cavitation phenomena and the behavior of bubbles in acoustic fields. What is more, the use of ultrasound in manufacturing technologies, such as intensification of hydrometallurgical processes in acoustic fields, ultrasonic cleaning and degassing, necessitate the development of nondestructive acoustic-field inspection methods. All the above processes and phenomena constitute a vast application field for nonperturbative laser techniques for measuring the parameters of acoustic and ultrasonic fields [6–9].

The optical density of a medium exposed to an acoustic field varies in space and time, the refractive index of the medium changing accordingly, depending on the parameters of the acoustic field and the acoustooptical properties of the medium. Consider how the refractive index of the medium is related to the parameters of the acoustic wave. Let a plane acoustic wave propagate in a medium along the x-axis at $z>0$. The variation of the current coordinate χ of a volume element of the medium is in this case described by the equation

$$\chi(x,t) = x - A\cos(\Omega_a t - K_a x), \tag{3.5}$$

the speed of the vibrational motion of the element being

$$v(x,t) = V\sin(\Omega_a t - K_a x), \tag{3.6}$$

where $A = V/\Omega_a$ is the vibration amplitude, V is the vibrational speed amplitude, Ω_a is the acoustic frequency, $K_a = 2\pi/\Lambda_a$ is the wave vector modulus, $\Lambda_a = c_a/\Omega_a$ is the length of the acoustic wave, and c_a is the velocity of sound in the medium. Note that relations (3.6) and (3.7) hold true at $V \ll c_a$.

The amplitude of variation of the relative density of the medium in the acoustic wave region, $S = \Delta\rho/\rho_0$, $\Delta\rho$ being the amplitude of density variations and ρ_0, the density of the unperturbed medium, is called the acoustic wave amplitude. This quantity determines the amplitudes of variation of:

$$\text{Refractive index,} \quad \frac{\Delta n}{n_0} = C_0 S, \tag{3.7}$$

$$\text{Acoustic pressure,} \quad \Delta p = p_0 \gamma S, \tag{3.8}$$

$$\text{Vibrational speed of the volume element,} \quad V = c_a S, \tag{3.9}$$

$$\text{Displacement of the volume element,} \quad A = \frac{\Lambda_a}{2\pi} S, \tag{3.10}$$

where C_0 is the material constant, γ is the compressibility coefficient, and n_0 is the refractive index of the unperturbed medium.

In the presence of acoustic wave, the dependence of the refractive index on co-ordinates and time has the form

$$n(x,t) = n_0 + \Delta n \sin (\Omega_a t - K_a x).$$

Generalizing this expression to the case where the acoustic wave propagates at an arbitrary angle of θ to the z-axis, we get

$$n(x,z,t) = n_0 + \Delta n \sin \Phi_a (x,z,t), \qquad (3.11)$$

where Δn is defined by expression (3.8) and $\Phi_a (x, z, t)$ is the phase of the acoustic wave at the point (x, z) at the instant t:

$$\Phi_a(x,z,t) = \Omega_a t - K_a x \sin \theta - K_a z \cos \theta. \qquad (3.12)$$

Expressions for the displacements of a particle within the volume element along the x- and z-axes have the respective forms

$$\chi (x,z,t) = x - A_x \cos \Phi_a (x,z,t), \qquad (3.13)$$

$$\eta (x,z,t) = z - A_z \cos \Phi_a (x,z,t), \qquad (3.14)$$

where $A_x = V_x/\Omega_a$ and $A_z = V_z/\Omega_a$ are the displacement amplitudes and $V_x = Sc_a \sin \theta$ and $V_z = Sc_a \cos \theta$ are the amplitudes of the vibrational speed projections.

Using the well-known expression $\gamma = c_a^2(\rho_0/p_0)$ [10], one can establish the relationship between the vibrational speed and acoustic pressure amplitudes:

$$\Delta p = c_a \rho_0 V. \qquad (3.15)$$

The variation of the refractive index is related to that of the acoustic pressure amplitude as follows:

$$\Delta n = \frac{C_0 n_0}{\gamma p_0} \Delta p. \qquad (3.16)$$

Table 3.3 lists as an example the main parameters of an acoustic wave and the correspondent refractive index variation amplitudes for two different media—water and air.

Table 3.3 Acoustic wave parameters in water and air for $f_a = 148$ kHz

Medium	Δp (Pa)	$\Delta n/\Delta n_0$	V (m/s)	A (m)	Λ_a (m)
Water	8.8×10^4	9.6×10^{-6}	0.059	6.4×10^{-8}	0.01
Air	5.6	1.2×10^{-8}	0.013	1.3×10^{-8}	0.0022

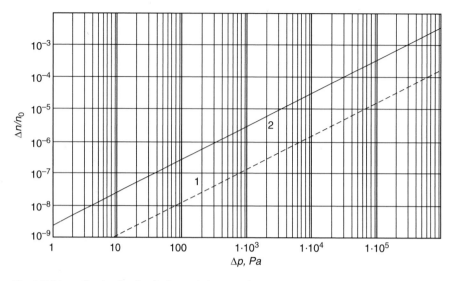

Fig. 3.7 Normalized refractive index variation amplitude as a function of the acoustic pressure for water (*2*) and air (*1*)

The values listed were calculated under the following conditions: the acoustic wave amplitude $S = 4 \times 10^{-5}$, acoustic wave frequency $f_a = 148$ kHz, atmospheric pressure $p_0 = 10^5$ Pa. The constants taken for water were $C_0 = 0.24$, $\gamma = 2.2 \times 10^4$, $c_a = 1485$ m/s, and $n_0 = 1.33$, and those for air, $C_0 = 2.9 \times 10^{-4}$, $\gamma = 1.4$, $c_a = 332$ m/s, and $n_0 = 1$

The relationship between the normalized refractive index variation amplitude and the acoustic pressure for air and water is shown in Fig. 3.7.

3.3   Liquid Mixing Processes

3.3.1   Characteristics of Processes Taking Place in the Course of Mixing of Liquid Media

Of importance in chemical technology is the development of methods for the optimization of manufacturing processes that convert the feed stock (raw materials) into finished products meeting the specified requirements [11].

There is a wide variety of processes making the feed stock undergo deep transformations accompanied by changes in its composition and state of aggregation. Several types of physical, physicochemical, and biological processes usually take their course simultaneously, entailing alterations in the optical characteristics of the medium.

The intensification of heat and mass exchange processes in process vessels is for the most part attained by the following methods [11, 12]:

- Increasing the specific contact surface area
- Raising the efficiency of mixing
- Improving the methods for implementing the contact of phases
- Increasing the relative velocity of phases
- Using nonstationary interphase exchange conditions providing for high instantaneous values of the heat and mass transfer coefficients
- Using nonequilibrium systems with high temperature and concentration gradients
- Running heat and mass transfer processes under unstable interface conditions

In most cases, the efficiency of mixing is the key factor that determines the efficiency of the entire process.

The analysis of the current status of the research on mixing processes in liquid media bears witness to the existence of the following main avenues of investigation:

- Study of flow structures, velocity fields, and degree of turbulence in process vessels equipped with mixing devices
- Development and investigation of new mixing methods; development of new and improvement of the existing mixing devices
- Mathematical description and modeling of mixing processes, including the solution of scaling problems
- Development of optical methods for determining the mixing uniformity and homogenization time

As mixing in liquid media is a hydrodynamic process, its analysis requires a detailed study of hydrodynamics in process vessels with mixing devices. In such vessels, there are formed complex three-dimensional flows wherein the motion of liquid depends not only on the type of the mixing device and vessel geometry used, but also on the rotational speed of the agitator and physicochemical properties of the liquid being mixed.

The need to produce the necessary hydrodynamic conditions in process vessels with mixing devices determines the main lines of inquiry into the structure of flows formed in process vessels equipped with various types of mixing device. These are as follows:

- Investigation of flow velocities and degree of turbulence in characteristic flow regions of the liquid being mixed
- Investigation of the uniformity of mixing
- Investigation of the homogenization time
- Investigation of the mixing time

All these tasks can be successfully performed with the aid of laser methods [3]. The hydrodynamic phenomena occurring in the course of liquid motion make for certain interactions among the components and volume elements of the interacting mixtures. The intensification of the mixing process and the extent of attainment of the desired technological result depend on the hydrodynamic liquid flow conditions. These conditions in turn depend on the geometry of the process vessel, the physical properties of the liquid, and the amount of energy deposited in a unit liquid volume.

The mixing process can be characterized by such general parameters as the specific power consumption, circulation capacity, and mixing time. The up-to-date methods for computing this process require knowledge of the velocity field in the process vessel, the degree of turbulence, the normal and tangential stresses, and the distribution of the concentration and temperature fields.

In process vessels with mixers, three-dimensional flows of liquid are formed whose motion depends not only on the vessel geometry, but also on the rotational speed of the agitator and the nature of the liquid.

3.3.2   Methods for Visualization of Twisted Liquid Flows

To study in detail the processes taking their course during operation of a process vessel with a mixing device, use is made of various optical methods for the visualization of liquid flows.

The visualization of twisted flows makes for a better understanding of the nature and mechanisms of the phenomena involved, and reveals and evaluates the influence of various hydrodynamic parameters and constructional details.

The optical homogeneity of the medium very often presents a problem in the study of the structure of such flows. This can be exemplified by water, aqueous glycerin solutions, and other transparent optically homogeneous liquids that are frequently used as model media in chemical engineering. To render the flow observable, it is necessary to make it visible. To this end, optical inhomogeneities in the form of fine particles are injected into the flow, or formed within it.

Visualization of liquid flows enables one to approach the solution of a very important problem—the formation of liquid flows with the desired parameters. When used in conjunction with recording instruments of appropriate characteristics, visualization of liquid flows can provide reliable information about the flow direction, velocity, and accelerations; the transition from laminar to turbulent flow; and the formation and decay of vortices.

Flow visualization data can serve as a starting material in creating mathematical models for computing heat and mass exchange characteristics on the basis of local values of heat and mass transfer coefficients.

As compared to other measurement means, visualization is an effective tool that makes it possible to:

- Establish flow patterns
- Take a systems approach to the study of the process, with due regard for various factors affecting the complex liquid flow
- Take quantitative measurements of the velocity field and the degree of turbulence

Fig. 3.8 Streamline pattern
in a glycerin-filled mixer free
from partitions

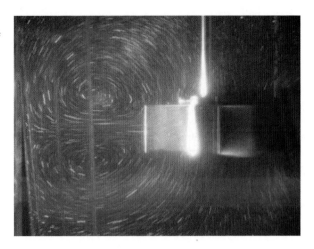

Figure 3.8 presents a typical mixing process visualization picture obtained in a mixer free from partitions and equipped with a centrally located four-blade agitator. To visualize streamlines in glycerin, use is made of aluminum powder and high-speed image acquisition.

One can see from this picture that the liquid motion in the mixer is of complex character: there are regions of intense mixing and also stagnation regions wherein the liquid is mixed but very weakly. All this necessitates the use of various flow research methods, namely, the Doppler method to determine local flow velocity characteristics, anemometry to determine the flow velocity field from particle images, and laser refractographic techniques to determine the uniformity of mixing and mixing time [3].

To select optimal visualization method to suit actual conditions presents certain problems. The choice requires that account should be taken of a number of factors, such as the scale of the process vessel under study, flow character and velocity, recording technique, and so on. For example, the scale of the phenomenon being studied places certain demands on the choice of both the visualization method and the recording and illuminating equipment. In most cases, the methods that are a success in the studies of edge effects cannot be used to investigate large-scale flows.

Temperature and concentration fluctuations in the course of mixing produce continuous refractive index gradients in the medium. The optical inhomogeneities caused by these gradients can be converted into illuminance variation on a screen or photofilm. Refractive index gradients can be detected by interference and refraction methods [3].

Laser refractography makes it possible to determine the mixing time of two liquids differing in optical properties [13–15], for the mixed liquids become optically homogeneous.

There are various criteria for the degree of homogeneity of mixtures. The degree of nonuniformity of mixing is usually determined from the ratio of the maximum difference between local concentration values to the difference between the initial concentration C_i and the final concentration C_f:

$$\Delta C_{max} = \frac{|C_{max} - C_{min}|}{|C_i - C_f|}.$$

The quantity ΔC_{max} depends on the mixing time. The form of this relation is governed by both the hydrodynamic mixing conditions and the geometrical parameters of the mixer. The mixing time of two liquids is at present fairly difficult to calculate theoretically.

In systems formed without any changes in the volume and polarizability of the components, the relationship between the refractive index of the mixture and its composition is almost linear:

$$n = n_1 V_1 + n_2 V_2, \tag{3.17}$$

where n, n_1, and n_2 are the refractive indices of the mixture and its individual components and V_1 and V_2 are the volume fraction of the components ($V_1 + V_2 = 1$).

Figure 3.9 presents the refractive index of aqueous solutions of some substances as a function of their concentration. These plots give an indication of refractive index fluctuations that can occur in solutions during the course of mixing and can be visualized by means of laser refractography. Some examples of such visualization are presented in Sect. 6.6.

It is well known that the spatial inhomogeneity of the properties of a flow leads to its optical inhomogeneity. In the case of nonextremal parameters of the medium, its refractive index is linearly related to its physical properties:

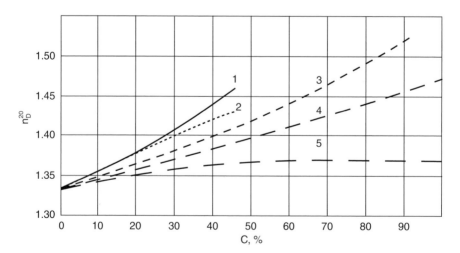

Fig. 3.9 Refractive index of aqueous solutions of some substances as a function of their concentration: *1*—sodium dichromate; *2*—NaCH; *3*—saccharose; *4*—glycerin; *5*—ethyl alcohol

$$n(S) = n(S_0) + (dn/dS)(S - S_0), \tag{3.18}$$

where $n(S)$ and $n(S_0)$ are the refractive indices of the medium with the parameters S and S_0, respectively, and dn/dS is the refractive index derivative with respect to the parameter S. If the flow parameters (temperature, density, pressure, salinity) depend on coordinates and time, so is the refractive index; i.e., $S(x,y,z,t) \to n[S(x,y,z,t)] \to n(x,y,z,t)$. When a laser plane passes through an optically inhomogeneous flow, refraction causes its individual portions to deflect and lose planarity, so that a complex curvilinear surface is formed that varies both in space and time.

Consider a case where the flow parameters and refractive index vary continuously along the x- and y-axes over a distance of l. We direct our laser plane along the x-axis. In that case, the angles α_x and α_y of deflection in the x- and y-directions of the laser plane element characterized by the initial coordinates (x_0, y_0) will be defined, at the exit from the flow, by the relations

$$\alpha_x(x,y,t) = \int_0^l \left(\frac{dn}{dS}\right)\left[\frac{dS(x,y,z,t)}{dx}\right] dz, \tag{3.19a}$$

$$\alpha_y(x,y,t) = \int_0^l \left(\frac{dn}{dS}\right)\left[\frac{dS(x,y,z,t)}{dy}\right] dz, \tag{3.19b}$$

where $dS(x,y,z,t)/dx$ and $dS(x,y,z,t)/dy$ are the gradients of the parameter S along the x and y-axes, respectively, that generally depend on all the spatial coordinates and the time t.

Since the refractive index gradient of a twisted flow under turbulent conditions is a random function, the trajectory of a laser beam in such a medium is also a random function, and, accordingly, its deflection angle at the exit from the medium will be random as well. Measured in the refraction method is the laser beam displacement observed on a semitransparent screen and recorded with a digital video camera.

In the course of mixing of two liquids differing in refractive index, for example, water and glycerin, the solution gradually gets homogenized, its refractive index becoming spatially equalized; i.e., the solution turns optically homogeneous. This causes the laser plane to regain its original form.

Use is often made of a solution of plain water with a common salt solution added. The homogenization time of such a solution is traditionally determined by measuring its conductivity with a conductometer whose remote transducer (probe) is immersed in the desired zone in the bulk of the solution. The shortcomings of this method are, first, the perturbation of the liquid flow because of the finite size of the measuring probe and secondly, the local character of measurements.

3.4   Hydrodynamic Phenomena in Stratified Liquids

Most widespread stratified liquid is sea water that is modeled in laboratory conditions by varying along the vertical coordinate the concentration of an aqueous sodium chloride solution [16]. The refractive index of such a solution depends on both temperature and salinity.

Figure 3.10 shows the refractive index of sea water as a function of salinity, and Fig. 3.11 that as a function of temperature [17]. One can see from the plots that in the most widespread temperature and salinity ranges the refractive index variation due to a 3°C change in temperature is of the same order of magnitude as that caused by a 1‰ change in salinity.

Hydrodynamic flows in stratified liquids induce intense mixing therein and as a result, bring about a small-scale optical inhomogeneity [3]. Figure 3.12 presents the results of visualization of degeneration of turbulence in a stratified liquid produced by varying along the vertical axis the concentration of an aqueous sodium chloride solution. The vortex-free turbulent region within the bulk of the stratified liquid was produced by operating a special turbulence stimulator for a fixed time. The evolution of the turbulent region depends on buoyancy forces, inertia, viscosity, and molecular diffusion. The shadowgram presented in Fig. 3.12a corresponds to the initial stage of development of the turbulent mixing region. Turbulence is observed in the form of a region of small-scale structures, and the waves generated upon the

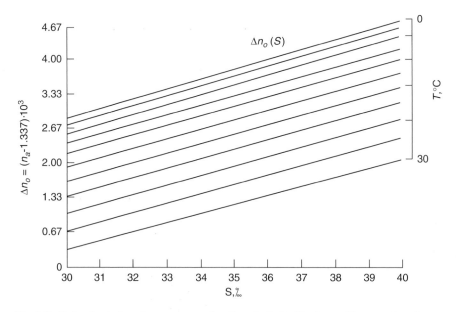

Fig. 3.10 Refractive index of sea water as a function of salinity. The interval between the adjacent curves corresponds to a 3°C change in temperature

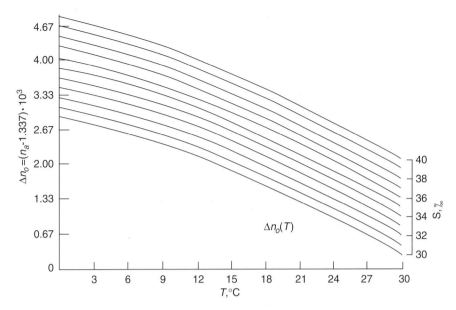

Fig. 3.11 Refractive index of sea water as a function of salinity. The interval between the adjacent curves corresponds to a 1‰ change in salinity

development of turbulence are seen as black-and-white strips. Figure 3.12b shows a shadowgram corresponding to a later stage of evolution of turbulence, wherein turbulence starts decaying at the boundary of the turbulent mixing region to form anisotropic structures seen as parallel strips on the periphery of the region.

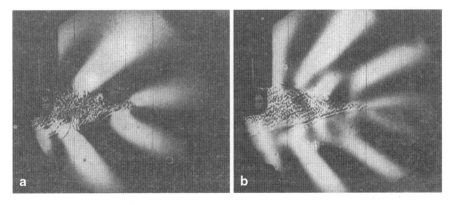

Fig. 3.12 Shadowgrams illustrating the evolution of a turbulent mixing region. **a** Early stage. **b** Late stage

References

1. W. Hauf and U. Grigull, *Optical Methods in Heat Transfer* (Academic Press, New York, 1970; Mir, Moscow, 1973).
2. M. V. Leikin, B. I. Molochnikov, V. N. Morozov, and E. S. Shakaryan, *Reflection Refractometry* (Mashinostroenie, Leningrad, 1983) [in Russian].
3. B. S. Rinkevichius, *Laser Diagnostics in Fluid Mechanics* (Begell House Inc. Publishers, New York, 1998).
4. L. D. Landau and E. M. Lifshits, *Theoretical Physics* (Nauka, Moscow, 1986), Vol. 6 [in Russian].
5. V. I. Artemov, O. A. Evtikhieva, K. M. Lapitsky *et al.*, "Numerical Modeling and Experimental Investigation of Free Convection in a Liquid around a Hot Body," in *Proceedings of the 4th Russian National Conference on Heat Exchange* (Moscow Power Engineering Institute Press, Moscow, 2006), Vol. 5, pp. 42–46 [in Russian].
6. V. A. Grechikhin, I. L. Raskovskaya, B. S. Rinkevichyus, and A. V. Tolkachev, "Investigation of the Acoustooptic Effect in the Interference Region of Laser Beams," *Kvantovaya Elektronika,* No. 8, 742 (2003).
7. I. L. Raskovskaya, "Laser Beam Propagation in a Medium with an Acoustic Wave," *Radiotekhnika i Elektronika,* No. 11, 1382 (2004).
8. I. L. Raskovskaya, B. S. Rinkevichyus, and A. V. Tolkachev, "Determination of the acoustic pressure in a liquid from the parameters of passing laser radiation," *Measurement Techniques,* **49**(6), pp. 605–609 (2006).
9. B. S. Rinkevichyus, O. A. Evtikhieva, and I. L. Raskovskaya, in CD *Proceedings International Colloquium on Physics of Shock Waves, Combustion, Detonation and Non-Equilibrium Processes, MIC 2005, Minsk, Nov. 14–19, 2005.*
10. A. Korpel, *Acoustooptics* (Marcel Dekker, New York and Basel, 1988).
11. V. G. Sister and Yu. V. Martynov, *Principles of Improving the Efficiency of Heat-Exchange Processes* (Bochkareva Publishing House, Kaluga, 1998) [in Russian].
12. V. A. Orlov and I. V. Chepura, "Mixing," in *Chemical Technology Processes and Apparatus. Transfer Phenomena, Macrokinetics, Similarity, Modeling, Designing,* Ed. by A. M. Kutepov (Logos, Moscow, 2001), Vol. 2 [in Russian].
13. A. Yu. Pokazentsev, P. A. Chaplina, and Yu. D. Chashechkin, *An Introduction to Ocean Optics* (Moscow State University Press, Moscow, 2007) [in Russian].
14. M. A. Bramson, E. I. Krasovsky, and B. V. Naumov, *Marine Refractometry* (Gidrometeoizdat, Leningrad, 1986) [in Russian].
15. M. V. Yesin, O. A. Evtikhieva, S. V. Orlov, B. S. Rinkevichyus, and A. V. Tolkachev, "Laser Refractometral Method for Visualization of Liquid Mixing in Twisted Flows," in CD Rom *Proceedings 10th International Symposium on Flow Visualization, Kyoto, Aug. 26–29, 2002.* Paper No. F037, pp. 1–8.
16. O. A. Evtikhieva, M. V. Yesin, S. V. Orlov, B. S. Rinkevichyus, and A. V. Tolkachev, "Laser Refraction Method for Studying Liquids in Twisted Flows," in *Proceedings of the 3rd Russian National Conference on Heat Exchage* (Moscow Power Engineering Institute, Moscow, 2002), Vol. 1, pp. 197–200 [in Russian].
17. O. A. Evtikhieva, A. I. Imshenetsky, B. S. Rinkevichus, and A. V. Tolkachev, "Visualization of Mixing in Twisted Flows by Means of Laser Planes," in CD ROM *Proceedings of the 2nd Russian Conference on Heat Exchange and Hydrodynamics in Twisted Flows* (Moscow Power Engineering Institute, Technical University, Moscow, 2005). Paper No. 0320500321.

Chapter 4
Refraction of Laser Beams in Layered Inhomogeneous Media

4.1 Geometrical Optics Approximation

4.1.1 Plane-Layered Medium

When the properties of the medium under study vary slowly enough as a function of coordinates, the propagation of laser beams can be described in terms of the geometrical optics approximation. In that case, the beam for structured laser radiation of any type should be represented in the form of a suitable family of rays. For layered inhomogeneous media [1, 2], one can obtain analytical expressions for ray trajectories and SLR projections on a screen. By a layered inhomogeneous medium we imply one whose refractive index depends on one coordinate (varies in one direction) only.

Let us derive equations for the trajectories of geometric-optical rays propagating in a layered inhomogeneous medium whose refractive index depends on a single rectangular coordinate; i.e., $n = n(x)$ (Fig. 4.1). The trajectory of the ray shown in Fig. 4.1 lies in the XOZ-plane and can be specified in the form of the function $z(x)$, $y = y_0 = $ const. The angle α is the one that the vector of the tangent to the ray makes with the x-axis (the direction of the tangent at a point in the trajectory $z(x)$ coincides with the ray direction at that point).

Obviously, the angle α will vary as a function of the position of the point in the ray trajectory, depending on the current coordinate x; i.e., $\alpha = \alpha(x)$. In the figure, $\alpha_0 = \alpha(0)$ and $n_0 = n(0)$. For the sake of definiteness, we take it that $n(x) = n_0$ throughout the region $x \leq 0$, which accords with the refraction problems considered later in the text.

To derive the trajectory equation, we will proceed from Snell's law for the plane inhomogeneous medium:

$$n(x) \sin \alpha(x) = n_0 \sin \alpha_0. \tag{4.1}$$

As follows from Fig. 4.1,

$$\tan \alpha(x) = \frac{dz}{dx}, \tag{4.2}$$

B. S. Rinkevichyus et al. (eds.), *Laser Refractography,*
DOI 10.1007/978-1-4419-7397-9_4, © Springer Science+Business Media, LLC 2010

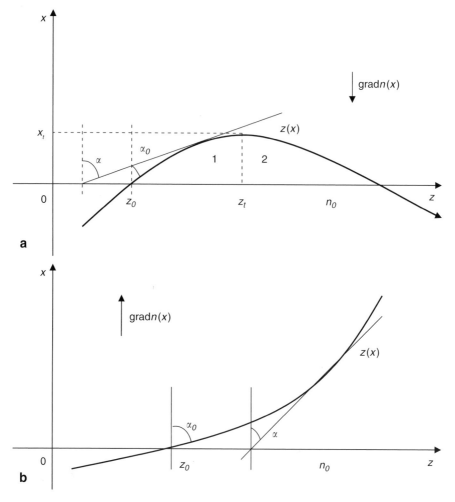

Fig. 4.1 Ray trajectories in a plane-layered medium. **a** The refractive index of the medium decreases. **b** The refractive index increases

and so, integrating (4.2), we get

$$z(x) = z_0 + \int_0^x \tan \alpha(x) \, dx, \qquad (4.3)$$

where $z_0 = z(0)$.

From formula (4.1) we obtain expression for $\tan \alpha(x)$ and substitute it into Eq. (4.3) to have

$$z(x) = z_0 + \int_0^x \frac{n_0 \sin \alpha_0 \, dx}{\pm\sqrt{n^2(x) - n_0^2 \sin^2 \alpha_0}}. \qquad (4.4)$$

The sign of the square root is determined by that of $\tan \alpha(x)$.

Relation (4.4) is the equation of the ray trajectory in a plane-layered medium, given the refractive index distribution $n(x)$ and the initial conditions $z_0 = z(0)$ and $\alpha_0 = \alpha(0)$.

The singularity in the integrand in expression (3.4) corresponds to the turning point x_t of the ray in the medium under analysis, which is defined such that

$$n(x_t) = n_0 \sin \alpha_0. \qquad (4.5)$$

Evidently a turning point can exist in the given medium only if grad $n(x) = (dn/dx) < 0$, which corresponds to the trajectory shown in Fig. 4.1a. If $dn/dx > 0$, then $n(x) < n_0$, and so equality (4.5) cannot be satisfied, irrespective of the value of x; i.e., there is no turning point in the medium (Fig. 4.1b).

When there is a turning point in the medium, trajectory (4.4) comprises two branches and is found as follows. First, we determine the coordinate $z_t = z(x_t)$ at the turning point as

$$z_t = z_0 + \int_0^{x_t} \frac{n_0 \sin \alpha_0 \, dx}{\sqrt{n^2(x) - n_0^2 \sin^2 \alpha_0}}, \qquad (4.6)$$

and then we find for $x < x_t$ the ascending branch of the trajectory (section 1 in Fig. 4.1a) to be

$$z_1(x) = z_0 + \int_0^{x} \frac{n_0 \sin \alpha_0 \, dx}{\sqrt{n^2(x) - n_0^2 \sin^2 \alpha_0}}. \qquad (4.7)$$

Next we find for $x > x_t$ the descending branch of the trajectory (section 2 in Fig. 4.1a) to be

$$z_2(x) = z_t + \int_{x_t}^{-\infty} \frac{n_0 \sin \alpha_0 \, dx}{-\sqrt{n^2(x) - n_0^2 \sin^2 \alpha_0}}.$$

or

$$z_2(x) = z_t + \int_{-\infty}^{x_t} \frac{n_0 \sin \alpha_0 \, dx}{\sqrt{n^2(x) - n_0^2 \sin^2 \alpha_0}}. \qquad (4.8)$$

Relations (4.5)–(4.8) form the basis for calculating the trajectories of geometric-optical rays in a plane-layered medium. Figure 4.2 presents as an example such trajectories for the exponential layer $n(x) = n_0\left(1 + \delta n e^{-\frac{x}{a}}\right)$, where δn is the maximum relative change of the refractive index and a is the characteristic size of the layer.

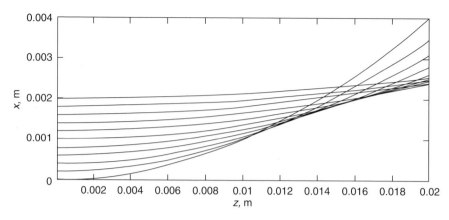

Fig. 4.2 Trajectories of geometric-optical rays in a plane exponential layer: $n_0 = 1.33$, $\delta n = 0.01$, $a = 1$ mm

4.1.2 Spherically Layered Medium

In a spherically layered medium, the refractive index n depends solely on the distance r to a fixed point that is convenient to superpose on the origin of coordinates. In that case, $\mathbf{r}$ is the radius vector of the observation point and $n = n(r)$. As demonstrated by Vinogradova and coworkers [1–3], the rays in this case are flat curves lying in a plane passing through the origin, with the condition

$$n(r)r\sin\alpha = \text{const},\tag{4.9}$$

being satisfied along each ray, where α is the angle between the tangent to the ray at the given point and the radius vector. Relation (4.9) is Snell's equation for spherically layered media.

To describe the ray trajectory, we use the spherical coordinates r, φ, θ (Fig. 4.3). For flat curves, the longitude $\varphi = \varphi_0 = \text{const}$, and so the trajectory can be described by the function $\theta(r)$, where θ is the angle between the radius vector and the positive direction of the z-axis, reckoned counterclockwise.

The following notation is introduced in Fig. 4.3: $\mathbf{r}$—the radius vector of a point on the ray, r—the radial coordinate of this point, r_t—the radial coordinate of the turning point, $\mathbf{k}$—the vector of the tangent to the ray at the point (r, θ), α—the angle between the vector of the tangent and the radius vector, γ—the angle between the vector of the tangent and the z-axis, and ρ—the impact parameter of the ray.

If the ray at an infinite distance from the origin of coordinates, prior to its entry into the inhomogeneous medium, is parallel to the z-axis, the impact parameter $\rho = \lim_{r\to\infty} r\sin\alpha(r)$, i.e., it is equal to the distance from the ray to the z-axis until the ray starts refracting in the inhomogeneity.

For a sufficiently high value of the radial coordinate r_0 at the entry to the medium, we may put $\rho = r_0\sin\alpha_0$, where $\alpha_0 = \alpha(r_0)$, and then we may rewrite condition (4.9) in the form

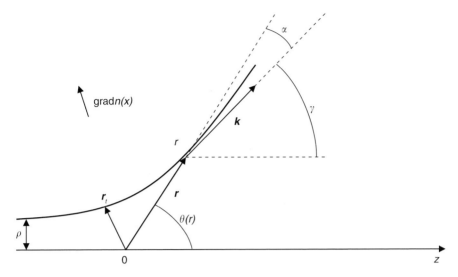

Fig. 4.3 Trajectory of a geometric-optical ray in a spherically layered medium

$$n(r)r \sin \alpha(r) = n_0 \rho, \tag{4.10}$$

where $n_0 = \lim_{r \to \infty} n(r)$, or $n_0 = n(r_0)$.

It follows from expression (4.10) that

$$\alpha(r) = \arcsin \frac{n_0 \rho}{rn(r)}. \tag{4.11}$$

To find the function $\theta(r)$ that defines the ray trajectory, we establish the relationship between the angles α and θ by using Fig. 4.4 showing an infinitely small section AB of the trajectory.

The radial coordinate at the point A is r, and the angle between the radius vector and the tangent is α. At the point B these quantities get incremented by dr and $d\alpha$, respectively. The dashed line in Fig. 4.4 describes a circle of radius r with its center at the origin of coordinates. The passage from the point A to the point B corresponds to the rotation of the radius vector through an angle of $d\theta < 0$.

Consider the curvilinear triangle ABC. Here $AC = -r \, d\theta$ and $BC = dr$, and so $\tan(\alpha + d\alpha) = -r \, d\theta/dr$. Disregarding the quantity $d\alpha$, we get

$$\tan(\alpha) = -\frac{r \, d\theta}{dr}. \tag{4.12}$$

From expression (4.11) we find

$$\tan \alpha = \frac{\sin \alpha}{\cos \alpha} = \frac{n_0 \rho}{\pm \sqrt{r^2 n^2(r) - \rho^2 n_0^2}}. \tag{4.13}$$

Fig. 4.4 Geometrical relationship between the angle functions $\alpha(r)$ and $\theta(r)$

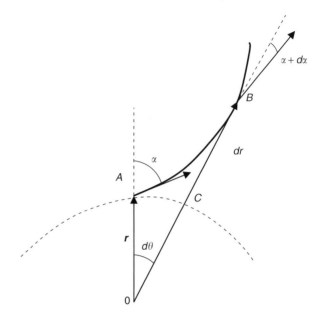

The sign of the square root depends on that of $\cos \alpha(x)$, which changes at the turning point r_t defined such that

$$r_t n(r_t) = n_0 \rho. \tag{4.14}$$

As in the case of plane-layered medium, the ray trajectory comprises two branches. For r varying from $r \to \infty$ at the entry to the medium to r_t at the turning point, the trajectory equation is defined by the relation following from expressions (4.12)–(4.13) at $\cos \alpha < 0$ (the subscripts 1 and 2 used here and elsewhere denote the pertinent variables before and after the turning point, respectively):

$$\frac{d\theta_1}{dr} = \frac{n_0 \rho}{r \sqrt{r^2 n^2(r) - \rho^2 n_0^2}}. \tag{4.15}$$

Integrating relation (4.15), we get

$$\theta_1(r) = \pi + \int_{\infty}^{r} \frac{n_0 \rho \, dr}{r \sqrt{r^2 n^2(r) - \rho^2 n_0^2}}, \tag{4.16}$$

where π is the value of the angle θ at the entry of the ray into the inhomogeneity.

It follows from expression (4.16) that the value of the angle θ at the turning point is

$$\theta_t = \pi - \int_{r_t}^{x} \frac{n_0 \rho \, dr}{r \sqrt{r^2 n^2(r) - \rho^2 n_0^2}}. \tag{4.17}$$

After the turning point we then have

$$\frac{d\theta_2}{dr} = -\frac{n_0\rho}{r\sqrt{r^2 n^2(r) - \rho^2 n_0^2}}. \tag{4.18}$$

Integrating expression (4.18), we obtain

$$\theta_2(r) = \theta_t - \int_{r_t}^{r} \frac{n_0\rho\, dr}{r\sqrt{r^2 n^2(r) - \rho^2 n_0^2}}. \tag{4.19}$$

Relations (4.16), (4.17), and (4.19) are directly used in computing refractograms for various types of structured laser radiation in spherically layered media (see Chap. 5).

Figure 4.5 shows ray trajectories in a spherical boundary layer formed in water around a hot ball of radius R, the refractive index inhomogeneity of the layer having a Gaussian profile of form $n(r) = n_0 - \Delta n e^{-\frac{(r-R)^2}{a^2}}$, where Δn is the refractive index variation, a is the thickness of the boundary layer, and n_0 is the refractive index of the unperturbed medium.

Of interest is to derive from relations (4.17) and (4.19) the asymptotic value of $\theta_2(r)$ at $r \to \infty$, which corresponds to the angle of deflection of the ray from its original direction:

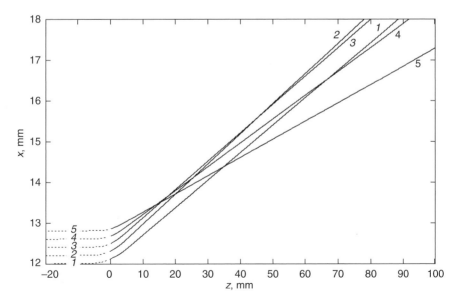

Fig. 4.5 Ray trajectories (*1* through *5*) in a spherical boundary layer with a Gaussian refractive index inhomogeneity profile: $n_0 = 1.33$, $\delta n = 0.01$, $a = 1$ mm, $R = 12$ mm

$$\lim_{r \to \infty} \theta_2(r) = \pi - 2 \int_{r_t}^{\infty} \frac{n_0 \rho \, dr}{r \sqrt{r^2 n^2(r) - \rho^2 n_0^2}}. \tag{4.20}$$

The ray deflection angle can also be obtained as the asymptotic value of the angle $\gamma(r)$ at $r \to \infty$.

It follows from Fig. 4.4 that $\gamma = \theta - \alpha$; then

$$\frac{d\gamma}{dr} = \frac{d\theta}{dr} - \frac{d\alpha}{dr}. \tag{4.21}$$

Since we are interested in the asymptotic value of the angle γ at the exit from the medium, use should be made of relation (4.18) for the derivative of θ_2—the second branch of the ray trajectory. We find the derivative of the angle α from relation (4.11) to be

$$\frac{d\alpha}{dr} = -\frac{n_0 \rho \left(\frac{1}{r} + \frac{n'(r)}{n(r)} \right)}{\sqrt{r^2 n^2(r) - \rho^2 n_0^2}}. \tag{4.22}$$

In that case,

$$\frac{d\gamma}{dr} = \frac{n_0 \rho n'(r)}{n(r) \sqrt{r^2 n^2(r) - \rho^2 n_0^2}}. \tag{4.23}$$

Integrating expression (4.23), we obtain the asymptotic ray deflection angle

$$\gamma = \gamma_t + \int_{r_t}^{\infty} \frac{n_0 \rho n'(r) dr}{n(r) \sqrt{n^2(r) r^2 - n_0^2 \rho^2}}. \tag{4.24}$$

It is evident from symmetry considerations that

$$\gamma_t = \int_{r_t}^{\infty} \frac{n_0 \rho n'(r) dr}{n(r) \sqrt{n^2(r) r^2 - n_0^2 \rho^2}}; \tag{4.25}$$

then

$$\gamma = \int_{r_t}^{\infty} \frac{2 n_0 \rho n'(r) dr}{n(r) \sqrt{n^2(r) r^2 - n_0^2 \rho^2}}. \tag{4.26}$$

In accordance with relations (4.20), (4.17), and (4.26), for $r \to \infty$ at the exit from the medium, we have

$$\gamma = \lim_{r \to \infty} \theta_2(r) = 2\theta_t - \pi. \tag{4.27}$$

Though the determination of the asymptotic deflection angles (4.26) of SLR rays fails to uniquely define the position of the refractogram on the screen, it allows the form of the refractogram to be described in a number of cases well enough.

Relations (4.16), (4.17), (4.19), and (4.26) obtained above describe the refraction of rays, but the classical geometrical optics approximation is inapplicable to the description of the diffraction effects associated with the spatial finiteness of laser beams and their diffraction by inhomogeneities. To evaluate and possibly allow for these wave effects, we hereinafter consider a quasioptical approximation to the description of the propagation of laser beams in weakly inhomogeneous media.

4.2 Quasioptical Approximation for Laser Beams in Weakly Inhomogeneous Media

Laser beams for any type of structured laser radiation can be represented in the form of a spatial (angular) spectrum determined by the given type of diffraction optical element. The propagation of each spatial harmonic in an inhomogeneous medium can be analyzed independently and then integrated in accordance with the superposition principle at the exit from the medium [4]. Analyzed below is a mathematical model of the propagation of a laser beam in an inhomogeneous medium, which accords with this approach.

Consider a laser beam propagating along the z-axis in a medium with a refractive index of n_0. Let $E(x, y, 0)$ be the complex field amplitude of the beam at the entry to the medium, i.e., at $z=0$. To find: the field $E(x, y, z)$ at the observation point. WE take the two-dimensional Fourier transform of the function $E(x, y, 0)$:

$$E(x,y,0) = \int\limits_{-\infty}^{+\infty}\int\limits_{-\infty}^{+\infty} F_0(k_x, k_y) \exp\left[j\left(k_x x + k_y y\right)\right] dk_x\, dk_y, \qquad (4.28)$$

where k_x and k_y are the components of the wave vector $\mathbf{k}$ and

$$F_0(k_x, k_y) = \frac{1}{(2\pi)^2} \int\limits_{-\infty}^{+\infty}\int\limits_{-\infty}^{+\infty} E(x,y,0) \exp\left[-j\left(k_x x + k_y y\right)\right] dx\, dy \qquad (4.29)$$

is the angular (spatial) spectrum of the function $E(x, y, 0)$. If the angular spectrum $F(k_x, k_y, z)$ is known, no matter what the value of z, the sought—for function $E(x, y, z)$ is defined by the expression [1]

$$E(x,y,z) = \int\limits_{-\infty}^{+\infty}\int\limits_{-\infty}^{+\infty} F(k_x, k_y, z) \exp\left[j\left(k_x x + k_y y\right)\right] dk_x\, dk_y. \qquad (4.30)$$

Since the function $E(x,y,z)$ satisfies the wave equation

$$\Delta E + k^2 E = 0, \tag{4.31}$$

then, substituting expression (4.30) into Eq. (4.31), we get the following differential equation for the function $F(k_x, k_y, z)$:

$$\frac{d^2 F}{dz^2} + \left(k^2 - k_x^2 - k_y^2\right) F = 0. \tag{4.32}$$

Solving this equation, subject to the condition $F(k_x, k_y, 0) = F_0(k_x, k_y)$, we find the partial solution corresponding to the wave propagating in the positive direction of the z-axis:

$$F_0(k_x, k_y, z) = F_0(k_x, k_y) \exp\left(jz\sqrt{k^2 - k_x^2 - k_y^2}\right). \tag{4.33}$$

As the observation point is moved farther away from the entrance to the medium, the angular spectrum changes as a result of the phase shift between different spectral components (plane waves propagating at different angles to the z-axis). Consider the propagation of a wave beam with a narrow angular spectrum; i.e., a beam whose width is much greater than the optical wavelength. This means that the transverse components k_x and k_y of the wave vector are small compared to its magnitude k. In that case, one can expand the expression $\sqrt{k^2 - k_x^2 - k_y^2}$ in the exponential of relation (4.33) into a series and retain only the terms quadratic in k_x and k_y. The beam will then be described, according to expression (4.30), by the function

$$E(x,y,z) = e^{ikz} \int\limits_{-\infty}^{+\infty}\int\limits_{-\infty}^{+\infty} F_0(k_x, k_y) \exp\left[j(k_x x + k_y y)\right]$$

$$\exp\left[-\frac{jz}{2k}\left(k_x^2 + k_y^2\right)\right] dk_x\, dk_y = e^{ikz} A(x,y,z), \tag{4.34}$$

where $A(x,y,z)$ is the wave amplitude and, according to relation (4.29),

$$F_0(k_x, k_y) = \frac{1}{(2\pi)^2} \int\limits_{-\infty}^{+\infty}\int\limits_{-\infty}^{+\infty} E(\xi, \eta, 0) \exp\left(-j(k_x\xi + k_y\eta)\right) d\xi\, d\eta. \tag{4.35}$$

Substituting relation (4.35) into (4.34) and integrating with respect to k_x and k_y, we obtain the following expression for the wave amplitude:

$$A(x,y,z) = \int\limits_{-\infty}^{+\infty}\int\limits_{-\infty}^{+\infty} G(x - \xi, y - \eta, z)\, A(\xi, \eta, z = 0)\, d\xi\, d\eta, \tag{4.36}$$

where the Green function

$$G = \frac{\exp\left(-\frac{(x-\xi)^2}{4\Lambda z}\right)}{\sqrt{4\pi \Lambda z}} \times \frac{\exp\left(-\frac{(y-\eta)^2}{4\Lambda z}\right)}{\sqrt{4\pi \Lambda z}}. \tag{4.37}$$

The analysis of the structure of expression (4.36) shows that it is the exact solution of the parabolic equation with the imaginary diffusion coefficient

$$\Lambda = -\frac{1}{2jk}, \tag{4.38}$$

which describes the propagation of the beam in the quasioptical approximation. The mathematical model and geometrical parameters of the problem are illustrated by Fig. 4.6. In the region $z \geq 0$, there is a plane or spherically layered inhomogeneous medium, the vector $\mathbf{K}_0$ specifying the direction of the field gradient of the inhomogeneity. The laser beam with an effective radius of w propagates in the XOZ-plane, $k = 2\pi/\lambda$ is the optical wave vector, where λ is the optical wavelength in the medium, and α is the angle the beam axis makes with the z-axis at $z=0$. Next, according to measurement conditions (see Chap. 6), we assume that $\sin \alpha \ll 1$.

Let $E(x, y, 0)$ be the complex amplitude of the beam field at the entry to the medium, $z=0$, whose refractive index at $z \geq 0$ may be represented in the form

$$n(x, y, z) = n_0 + \Delta n(x, y, z), \tag{4.39}$$

the relation $\delta n = \frac{\Delta n}{n_0} \ll 1$ being satisfied, where n_0 is the refractive index of the unperturbed medium and Δn, the maximum deviation from n_0. To find: the complex amplitude $E(x, y, z)$ in the medium at the observation point $P(x, y, z)$.

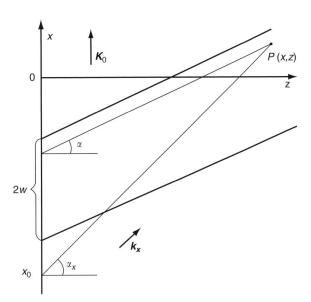

Fig. 4.6 Geometrical parameters of the problem

To solve the problem posed, the beam field is represented in the form of a spatial spectrum and the propagation of each spectral component in the inhomogeneous medium is described in terms of the geometrical optics approximation, using the eikonal and amplitude perturbation methods [2]. The optical field at the observation point is a superposition of partial beams whose interference, with due regard given to perturbations, distorts the beam amplitude and phase. The application of the notion of geometric-optical rays to the partial waves imposes the following limitation on the distance z, given the wavelength λ and characteristic size a of the inhomogeneity:

$$\frac{\lambda z}{a^2} \ll 1. \tag{4.40}$$

We represent the complex amplitude of the optical field at $z=0$ on the form

$$E(x,y,0) = \exp\{jkx \sin \alpha\} A(x,y,0), \tag{4.41}$$

where $A(x,y,0)$ is the complex field amplitude at $\alpha = 0$. We resolve $E(x,y,0)$ into a spectrum of plane waves with the parameters k_x and k_y:

$$E(x,y,0) = \exp\{jkx \sin \alpha\} \int_{-\infty}^{\infty} \int_{-\infty}^{\infty} F(k_x, k_y) \exp\{j(k_x x + k_y y)\} dk_x \, dk_y, \tag{4.42}$$

where $F(k_x, k_y)$ are the complex amplitudes of the spectral components,

$$F(k_x, k_y) = \frac{1}{(2\pi)^2} \int_{-\infty}^{\infty} \int_{-\infty}^{\infty} A(x,y,0) \exp\{-j(k_x x + k_y y)\} dx \, dy, \tag{4.43}$$

and the propagation direction of the pertinent plane waves is characterized by the vector with the components $\left(k_x + k \sin \alpha, k_y, \sqrt{k^2 - (k_x + k \sin \alpha)^2 - k_y^2}\right)$. In the XOZ-plane in Fig. 4.6 a geometric-optical ray is shown making an angle of α_x with the z-axis and corresponding to the partial plane wave with the parameters

$$k_x = k \sin \alpha_x - k \sin \alpha, \quad k_y = 0. \tag{4.44}$$

In accordance with [2], the phase of each spectral component at the observation point $P(x, y, z)$ may be represented in the form:

$$\varphi(x,y,z,k_x,k_y) = (k_x + k \sin \alpha)x + k_y y$$
$$+ z\sqrt{k^2 - (k_x + k \sin \alpha)^2 - k_y^2} + \Delta\varphi \tag{4.45}$$

and is found by integrating the values of refractive index (4.39) along the appropriate geometric-optical ray. The first three terms on the right-hand side of expression

(4.45) correspond to the phase of the partial wave in the unperturbed medium, and the last term is the phase perturbation due to the optical inhomogeneity, defined as

$$\Delta\varphi(x, y, z, k_x, k_y) = \delta n \cdot f(x, y, z, k_x, k_y). \tag{4.46}$$

The function $f(x, y, z, k_x, k_y)$, hereinafter designated f for brevity, depends on the structure of the actual inhomogeneity (see the example cited later in the text).

The field at the observation point $P(x, y, z)$ is a superposition of partial waves, with their propagation conditions taken into account:

$$E(x, y, z) = \int_{-\infty}^{+\infty} \int_{-\infty}^{+\infty} \frac{F(k_x, k_y)}{\sqrt{\gamma(k_x, k_y, x, y, z)}} \exp\{j[(k_x + k\sin\alpha)x + k_y y$$
$$+ z\sqrt{k^2 - (k_x + k\sin\alpha)^2 - k_y^2} + \Delta\varphi]\} dk_x \, dk_y, \tag{4.47}$$

where $\gamma(k_x, k_y, x, y, z)$ allows for the divergence of the rays in the inhomogeneous medium and is found on the basis of the transfer equations [2] for each spectral component.

For beams with a narrow spatial spectrum ($\lambda/w \ll 1$), further simplifications are possible. Based on the work by Raskovskaya [4], the representation of the complex amplitude of the laser beam at the observation point has the following form, accurate up to the expansion terms of the functions in the quantity in exponent (4.47) that are quadratic in k_x:

$$E(x, y, z) = \frac{\exp\{j[kz\cos\alpha + kx\sin\alpha + \Delta\varphi_0]\}}{\sqrt{\gamma(0, 0, x, y, z)}} A(x, y, z), \tag{4.48}$$

where

$$\Delta\varphi_0 \equiv \Delta\varphi_0(x, y, z) \equiv \delta n \cdot f_0(x, y, z), \tag{4.49}$$

$$A(x, y, z) = \int_{-\infty}^{+\infty} \int_{-\infty}^{+\infty} F(k_x, k_y) \exp\{j[k_x(x - z\tan\alpha$$
$$+ \delta n f') + k_y y - \frac{k_x^2}{2k}(z - \delta n k f'') - \frac{k_y^2}{2k} z]\} dk_x \, dk_y, \tag{4.50}$$

and f' and f'' are partial derivatives with respect to k_x.

The first term in expression (4.50) corresponds to the field of the plane wave propagating in the inhomogeneous medium at an angle of α to the z-axis. The function $\Delta\varphi_0(x, y, z)$ in the quantity in exponent specifies the phase perturbation along the beam arriving at the observation point (Fig. 4.6), and the divergence function $\gamma(0, 0, x, y, z)$, which will be defined for the given inhomogeneity configuration elsewhere in the text, describes to the zeroth approximation the diffraction effects due to the spatial inhomogeneity with a characteristic size of a.

The second term, $A(x, y, z)$, allows for the effects associated with the spatial limitation of the beam; i.e., its diffraction in conditions of inhomogeneous medium, caused by the existence of the characteristic size w.

To obtain a convenient analytical expression describing these effects, we compare expressions (4.48) with the following expression for the complex amplitude $E^0(x, y, z)$ of the optical field at the observation point $P(x, y, z)$ in the unperturbed medium; i.e., at $\delta n = 0$:

$$E^0(x, y, z) = \exp\{j[kz \cos\alpha + kx \sin\alpha]\} A^0(x, y, z), \qquad (4.51)$$

$$A^0(x, y, z) = \int\limits_{-\infty}^{\infty}\int\limits_{-\infty}^{\infty} F(k_x, k_y) \exp\{j[k_x(x - z \tan\alpha)$$

$$+ k_y y - \frac{k_x^2}{2k}z - \frac{k_y^2}{2k}z]\} dk_x\, dk_y. \qquad (4.52)$$

If expression (4.52) allows separating the variables, for example, for Gaussian beams

$$A^0(x, y, z) = A_x^0(x, z) A_y^0(y, z), \qquad (4.53)$$

then

$$A(x, y, z) = A_x^0(x + \delta n f', z - \delta n k f'') A_y^0(y, z). \qquad (4.54)$$

Considering that the function f and its partial derivatives with respect to k_x depend on the coordinates of the observation point $P(x, y, z)$ and introducing the notation

$$\Delta x(x, y, z) = \delta n f', \qquad \Delta z(x, y, z) = -\delta n k f'', \qquad (4.55)$$

we finally write down expression (4.48) in the form

$$E(x, y, z) = \frac{\exp\{j[kz \cos\alpha + kx \sin\alpha + \Delta\varphi_0(x, y, z)]\}}{\sqrt{\gamma(x, y, z)}}$$

$$\times A_x^0(x + \Delta x(x, y, z), z + \Delta z(x, y, z)) A_y^0(y, z). \qquad (4.56)$$

Expression (4.56) allows one to find at the observation point $P(x, y, z)$ the complex amplitude of a laser beam propagating in a weakly inhomogeneous medium, given its complex amplitude in the homogeneous medium. The complex amplitude of a beam in a medium perturbed, for example, by a temperature field, is expressed in terms of the complex amplitude of the beam in the unperturbed medium at the same observation point by formally making the substitution of coordinates $x \rightarrow x + \Delta x$ and $z \rightarrow z + \Delta z$ in the function A_x^0, where the function $\Delta x = \Delta x(x, y, z)$ describes the distortions of the complex amplitude of the beam associated with the refraction-induced shift of rays along the z-axis within the beam cross-section, and $\Delta z = \Delta z(x, y, z)$ specifies the change in the conditions of focusing and diffraction-induced broadening of the beam in the inhomogeneous medium.

The wave description of the field of beams allows one to take account of the diffraction effects occurring in the studies of optically inhomogeneous media by the laser refractography techniques.

4.3 Numerical Modeling of the Propagation of a Beam in a Weak Exponential Temperature Inhomogeneity

To determine the functions $\Delta \varphi_0(x, y, z)$, $\Delta x(x, y, z)$ and $\gamma(x, y, z)$ defined by formulas (4.49) and (4.55), we render expression (4.39) concrete by assuming that the inhomogeneity of the medium is formed by a temperature field, the temperature dependence of the refractive index is given by relation (3.3), and the beam propagation angle is $\alpha = 0$. To reveal the fundamental physical laws governing the propagation of the beam, consider the exponential coordinate dependence of the refractive index:

$$n(x, t) = n_0 \left(1 + \delta n e^{-\frac{x}{a}} \right), \tag{4.57}$$

where a is the characteristic thickness of the temperature inhomogeneity layer near a hot body.

The phase, according to expression (4.45), that the partial wave with the parameters $k_x = k \sin \alpha_x$, $k_y = 0$ has at the point (x, y, z) is found by integrating along the appropriate ray:

$$\varphi(x, z) = k \left\{ x_0(x, z) \sin \alpha_x + \int_{x_0}^{x} \left(1 + \delta n e^{-\frac{x}{a}} \right) \frac{dx}{\sin \alpha_x} \right\}, \tag{4.58}$$

where $x_0 = x - z \tan \alpha_x$ is the coordinate of the ray's entry into the medium (Fig. 4.6). Based on expressions (4.46), (4.47), and (4.58), we determine the function f characterizing the structure of the inhomogeneity:

$$f = kae^{-\frac{x}{a}} \left(\frac{e^{\frac{z \cdot \tan \alpha_x}{a}} - 1}{\sin \alpha_x} \right). \tag{4.59}$$

We expand function (4.47) into a series in k_x at the point $k_x = 0$ and use relations (4.49) and (4.55) with a view to finding explicit expressions for $\Delta \varphi_0(x, z)$, $\Delta x(x, z)$, $\gamma(x, z)$:

$$\Delta\varphi_0(x,z) = \delta n \cdot kz \cdot e^{\frac{-x}{a}}, \tag{4.60}$$

$$\Delta x(x,z) = \delta n \cdot \frac{z^2}{a} \cdot e^{\frac{-x}{a}}, \tag{4.61}$$

$$\gamma(x,z) = \sqrt{1 - \delta n \frac{z^2}{a^2} e^{-\frac{x}{a}}}. \tag{4.62}$$

Relations (4.56) and (4.59) through (4.62) define the optical field of the laser beam in the medium in the presence of temperature inhomogeneity (4.57).

We will perform numerical modeling for a Gaussian beam whose complex amplitude $A^0(x,y,z)$ in the homogeneous medium is defined by the expression

$$\begin{aligned} A^0(x,y,z) = \frac{A^0(0,0,z_F)}{\sqrt{1+D^2}} \exp\Bigg\{ &-\frac{x^2}{w^2(1+D^2)} \\ &-\frac{y^2}{w^2(1+D^2)} + j\psi(x,y,z)\Bigg\} \end{aligned} \tag{4.63}$$

where the function ψ specifies the shape of the wave front, D is the dimensionless diffraction length, and z_F is the coordinate of the waist:

$$\psi(x,y,z) = \frac{x^2+y^2}{w^2(1+D^2)} \cdot \frac{D}{1+D^2} - \arctan D, \tag{4.64}$$

$$D = \frac{(z-z_F)}{R_0}, \quad R_0 = \frac{kw^2}{2}, \tag{4.65}$$

where R_0 is the confocal parameter of the beam. Based on formulas (4.56) and (4.59) through (4.62), we computed in the first approximation in the eikonal and amplitude perturbation method the amplitude of the Gaussian beam at a distance of z to be

$$\begin{aligned} A(x,y,z) = &\frac{A^0(0,0,z_F)}{\sqrt{1+D^2 - \delta n \cdot \frac{z^2}{a^2} \cdot e^{\frac{-x}{a}}}} \\ &\times \exp\Bigg\{-\frac{\left(x - \delta n \cdot \frac{z^2}{a} \cdot e^{\frac{-x}{a}}\right)^2}{w^2(1+D^2)} - \frac{y^2}{w^2(1+D^2)}\Bigg\}. \end{aligned} \tag{4.66}$$

As follows from the structure of expression (4.66), the diffraction-induced divergence of the beam, owing to the limitation of its radius, is offset by its refraction-induced focusing (the terms in the radicand differ in sign), which corresponds to the

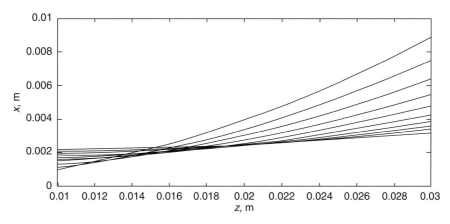

Fig. 4.7 Refraction of rays in a beam for a plane-layered medium model at $\delta n = 0.01$, $a = 1$ mm

growth of its amplitude. The first term in the quantity in exponent determines the displacement of the beam center; i.e., it describes diffraction in the wave interpretation, with allowance made simultaneously for the distortion of the shape of the beam envelope. The denominator going to zero in expression (4.66) corresponds to the formation of a caustic, which can be seen at $z > 1$ cm in Fig. 4.7 presenting a refraction pattern of rays in a beam. The computation was made for $\delta n = 0.01$ and $a = 1$ mm.

Figure 4.8 illustrates the displacement and deformation of a beam as a function of the distance z traveled in a spherical layer 0.3 mm distant from a hot ball 40 mm in diameter.

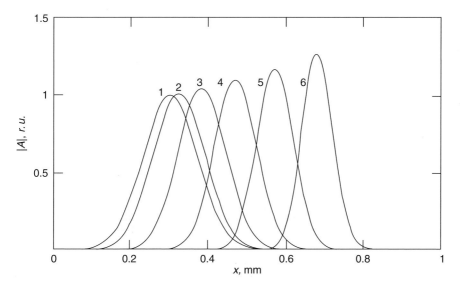

Fig. 4.8 Displacement and deformation of a beam as a function of the distance z: *1*—$z = 0$ mm; *2*—$z = 2$ mm; *3*—$z = 4$ mm; *4*—$z = 6$ mm; *5*—$z = 8$ mm; *6*—$z = 10$ mm

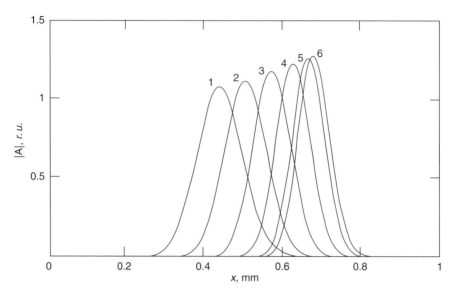

Fig. 4.9 Shape of a beam as a function of its lateral displacement: 1—$x_0 = 0.5$ mm; 2—$x_0 = 0.46$ mm; 3—$x_0 = 0.42$ mm; 4—$x_0 = 0.38$ mm; 5—$x_0 = 0.34$ mm; 6—$x_0 = 0.3$ mm

Analyzed at the same problem parameters in Fig. 4.9 is the shape of the beam as a function of the lateral displacement x_0 of its axis in the layer (in the radial direction). For a laser plane, x_0 is the distance from the surface of the hot ball to the center of the plane. As the beam approaches the center of the inhomogeneity, its displacement and focusing increase.

Figures 4.10a–c illustrate the focusing and distortion of the laser beam in the spherical layer as a function of its distance from the center of the inhomogeneity, the problem parameters remaining the same.

In this section, we have investigated the propagation of laser beams in layered inhomogeneities on the basis of wave equations, with due regard given to refraction in the first approximation in the method of perturbations in the parameter δn. The computation of laser plane propagation at various distances from the hot body at hand in the quasioptical approximation enables one to simultaneously take account of the diffraction and refraction effects in the processing of images using the method described in Chap. 8, which makes it possible to reduce the error of measurement of the laser plane displacement. What is more, the use of the wave description helps establish the applicability limits of geometrical optics; specifically, as follows for the above computations, the refraction effects predominate, and this allows using the classical ray methods. As shown by our numerical analysis, for temperature inhomogeneities of some 0.1–1 mm in size in water at a temperature difference of a few degrees centigrade, the diffraction effects due to the finite size of the beam are insignificant in comparison with their refraction counterparts, for in expression (4.66) $D^2 \ll \delta n \cdot z^2 / a^2$. For this reason, stress in the theoretical sections is on the study of refraction on the basis of ray equations in spherically inhomogeneous media.

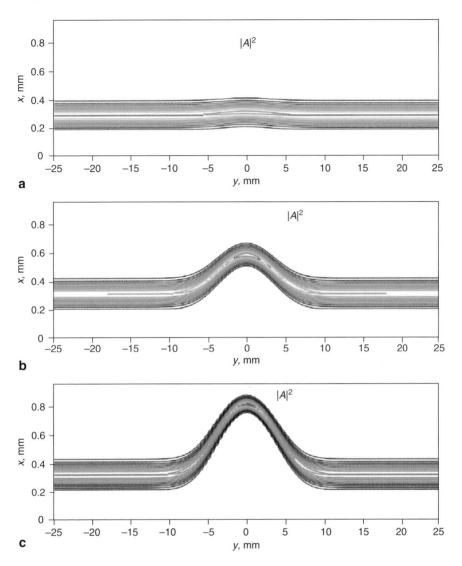

Fig. 4.10 Focusing and distortion of the laser beam in the spherical layer as a function of its distance from the center of the inhomogeneity. **a** $z = 2$ mm. **b** $z = 8$ mm. **c** $z = 12$ mm

References

1. M. B. Vinogradova, O. V. Rudenko, and A. P. Sukhorukov, *Wave Theory* (Nauka, Moscow, 1979) [in Russian].
2. Yu. A. Kravtsov and Yu. I. Orlov, *Geometrical Optics of Inhomogeneous Media* (Nauka, Moscow, 1980) [in Russian].
3. E. G. Zelkin and R. A. Petrova, *Lens Antennas* (Sovetskoe Radio, Moscow, 1974) [in Russian].
4. I. L. Raskovskaya, "Propagation of a Laser Beam in a Medium with an Acoustic Wave," *Radiotekhnika i Elektronika*, No. 11, 1382 (2004).

Chapter 5
Refraction of Structured Laser Radiation in Spherical Temperature Inhomogeneities

5.1 Refractograms of Plane Structured Laser Radiation for Typical Spherical Inhomogeneities

5.1.1 Refraction of Plane Structured Laser Radiation in a Spherical Inhomogeneity

To develop methods for constructing theoretical refractograms, we study the refraction of a laser plane in a spherically layered medium whose refractive index varies in accordance with expression (4.1).

The geometry of the problem is illustrated in Fig. 5.1. Laser plane *1* normal to the *x*-axis propagates along the *z*-axis. The origin of coordinates is superimposed on the center of hot ball *2*. The laser plane projection is observed in plane *3* normal to the *z*-axis that is placed at a distance of z_1 from the origin.

The connection between rectangular and spherical coordinates is specified by the relations

$$
\begin{aligned}
x &= r \sin \theta \cos \varphi, \\
y &= r \sin \theta \sin \varphi, \\
z &= r \cos \theta.
\end{aligned}
\tag{5.1}
$$

At $z = z_0$ the equation of the laser plane has the form $x = x_0$. The impact parameter of a ray lying in this plane depends on the parameter φ and is given by

$$
\rho(\varphi) = \frac{x_0}{\cos \varphi}.
\tag{5.2}
$$

In the case of spherical symmetry, this ray remains in the plane specified by the parameter φ, and its trajectory is defined by the function $r(\theta, \varphi)$. Ray trajectories are calculated on the basis of the relations describing the propagation of rays in spherically layered inhomogeneities [1] that are presented in Sect. 4.1.2.

The radial coordinate of the ray at its entry into the medium (at $z = z_0$) is

$$
r_0(\varphi) = \sqrt{\rho^2(\varphi) + z_0^2}.
\tag{5.3}
$$

B. S. Rinkevichyus et al. (eds.), *Laser Refractography*,
DOI 10.1007/978-1-4419-7397-9_5, © Springer Science+Business Media, LLC 2010

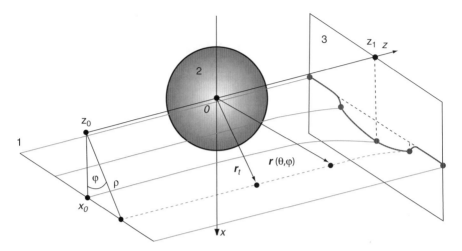

Fig. 5.1 Geometrical parameters in laser plane refraction: *1*—laser plane, *2*—hot ball, *3*—semi-transparent opaque screen

The angle θ_0 characterizes the direction of the ray at the entry into the medium:

$$\theta_0(\varphi) = \frac{\pi}{2} + \arctan \frac{z_0}{\rho(\varphi)}. \tag{5.4}$$

The angle $\theta = \theta_t$ corresponding to the turning point of the ray is given by

$$\theta_t(\varphi) = \theta_0(\varphi) + \int_{r_0(\varphi)}^{r_t(\varphi)} \frac{n_0 \rho(\varphi)\,dr}{r\sqrt{n^2(r)r^2 - n_0^2 \rho^2(\varphi)}}. \tag{5.5}$$

The ray equation before the turning point is

$$\theta(r, \varphi) = \theta_0(\varphi) + \int_{r_0(\varphi)}^{r} \frac{n_0 \rho(\varphi)\,dr}{r\sqrt{n^2(r)r^2 - n_0^2 \rho^2(\varphi)}}. \tag{5.6}$$

The ray equation after the turning point is

$$\theta(r, \varphi) = \theta_t(\varphi) + \int_{r}^{r_t(\varphi)} \frac{n_0 \rho(\varphi)\,dr}{r\sqrt{n^2(r)r^2 - n_0^2 \rho(\varphi)}}. \tag{5.7}$$

Formulas (5.5)–(5.7) form the basis for computing the refraction of laser planes in radial inhomogeneities. The angle φ is the parameter specifying an arbitrary ray in the laser plane, which makes it possible to describe the entire family of rays pertaining to this plane. On the screen located at a distance of z_1 from the origin, the radial coordinate $r(z_1, \varphi)$ is found from the equation

$$r \cos \theta(r, \varphi) = z_1, \tag{5.8}$$

and the coordinates of the laser plane projection on the screen are given by

$$\begin{aligned} x(z, \varphi) &= r(z, \varphi) \sin \theta(r(z, \varphi), \varphi) \cos \varphi, \\ y(z, \varphi) &= r(z, \varphi) \sin \theta(r(z, \varphi), \varphi) \sin \varphi. \end{aligned} \tag{5.9}$$

Relations (5.9) define the structure of the refractogram for the given inhomogeneity.

5.1.2 Boundary Layer Near a Hot Ball

As stated earlier, three-dimensional refractograms allow one to directly visualize and qualitatively diagnose inhomogeneities. But their quantitative diagnostics (Chap. 8) requires comparison between theoretical and experimental refractograms, which is more convenient to make using two-dimensional SLR projections (two-dimensional refractograms) in sections specified at certain distances from the inhomogeneity of interest. Therefore, this chapter is devoted to the methods of calculating two-dimensional refractograms, using as an example spherical temperature inhomogeneities [2–4].

Numerical calculations were made by formulas (5.7)–(5.9) for such parameters of the radial dependence of temperature, (3.4), as are close to the conditions of experiments with a hot ball immersed in water, described in Chap. 6. Figure 5.2 presents radial dependence curves for two models of temperature distribution near a hot ball, the model parameters being as follows:

1. $R = 12$ mm, $T_0 = 20°C$, $\Delta T = 70°C$, $\Delta R = 0$, $a = 1$ mm
2. $R = 12$ mm, $T_0 = 19.7°C$, $\Delta T = 80°C$, $\Delta R = -5$ mm, $a = 1.4$ mm

The equality $\Delta R = 0$ corresponds to the case where grad($T(R)$) = 0; and as one moves away from the surface of the ball, the temperature gradient first grows higher and then decreases, which is due to the presence of a turning point in curve 1. The situation in this case is such that the laser plane rays passing closer to the ball surface suffer weaker deflection than their counterparts passing farther away. Curve 2 in Fig. 5.2 corresponds to the second temperature distribution model (grad($T(R)$)≠0). In that case, there is no boundary layer of a relatively slow temperature variation. Curves 1a and 2a demonstrate the variation of the temperature gradient as a function of the radial coordinate for the first and the second model, respectively.

When a family of rays with the parameter $x_0 = 12.05$ mm refracted on an inhomogeneity of type 1 is projected at a distance of $z_1 = 155$ mm, local extrema are observed on the projection plot $x(y)$ (curve 1 in Fig. 5.3), which is caused by the above-described specificity of refraction of the rays.

The presence of extrema on the $x(y)$ plot corresponding to a projection of a refracted laser plane therefore points to the existence of a boundary layer with

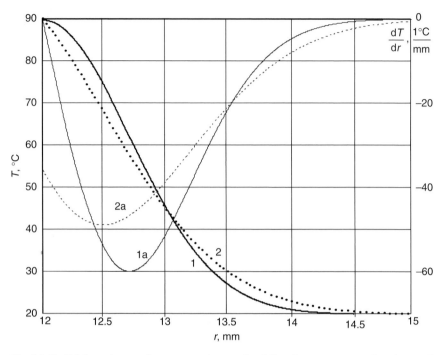

Fig. 5.2 Radial dependences of temperature (curves *1* and *2*) and temperature gradient (curves *1a* and *2a*) for two boundary layer models: *1*—grad($T(R)$) = 0, *2*—grad($T(R)$) ≠ 0

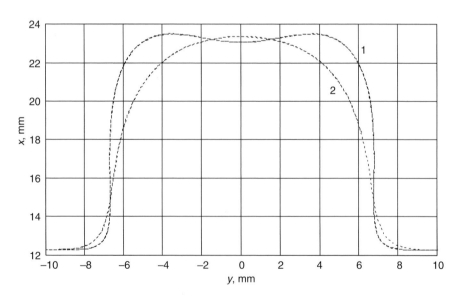

Fig. 5.3 Laser plane projections for the first (curve *1*) and the second (curve *2*) boundary layer model at $z_1 = 155$ mm, $x_0 = 12.05$ mm

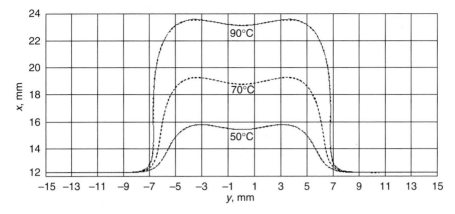

Fig. 5.4 Laser plane projections for type 1 boundary layer model at various ball surface temperatures. $R = 12$ mm, $z = 155$ mm, $x_0 = 12.05$ mm

a relatively slow temperature variation and $\mathrm{grad}(T(R)) = 0$. In the case of type 2 inhomogeneity, no local extrema appear for a family of rays with the parameter $x_0 = 12.05$ mm (curve *2* in Fig. 5.3), which allows one to conclude, when solving the inverse problem, that there is no boundary layer with zero temperature gradient.

Figure 5.4 presents theoretical laser plane projections that can illustrate the cooling of the ball.

Figure 5.5 shows typical 3D refractograms of a hot spherical layer in water, obtained with plane structured laser radiation (the z-axis scale is substantially reduced), and Fig. 5.6 presents a set of cross-sections of a 3D refractogram of the boundary layer near a hot body of spherical shape.

If there is an inhomogeneity region wherein the temperature gradient is relatively small or altogether absent, in the cross-sections of the refractogram local

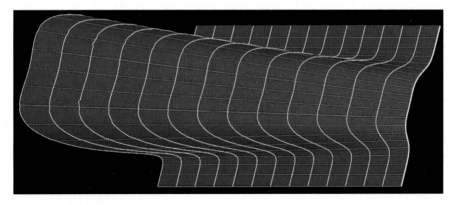

Fig. 5.5 Typical 3D refractogram of the boundary layer near a hot body of spherical shape. The *red lines* indicate the generating rays

Fig. 5.6 Set of cross-sections
of a 3D refractogram of the
boundary layer near a hot
body of spherical shape

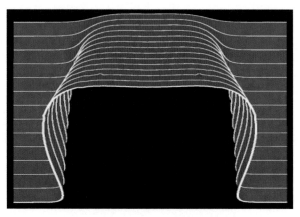

extrema due to the reduction of the laser beam deflection in low-gradient regions
appear weakly expressed (at actual temperatures) (Fig. 5.6).

The above analysis allows one to judge the character of temperature variation
in the boundary layer in the immediate vicinity of a hot ball in a liquid from the
specific features of the projection of a laser plane refracted in the temperature field.

Refraction was numerically calculated on the basis of a spherically layered medium model. The choice of this model was substantiated by the experimental fact
that the deflection of the ray takes place in a thin boundary layer directly beneath
the hot ball in the liquid. The ray practically reflects from this layer; i.e., the radius
of curvature of its trajectory is very small, and so is the region wherein the ray trajectory is other than rectilinear. For this reason, it suffices to know the temperature
field in this region important for refraction, namely, the thin layer directly beneath
the sphere. Theoretical temperature values in this region can be approximated (accurate to within a few hundredths fractions of a degree) by a radial dependence
corresponding to a spherically layered medium model.

The choice of such a model obviates the need for the thermophysical calculations
requiring substantial computational resources that are used to solve in the general
case 3D refraction problems, given the refractive index and its gradient throughout
the volume under study.

5.1.3 Boundary Layer Near a Cold Ball

Numerical calculations were made by formulas (5.7)–(5.9) for such parameters of
the radial dependence of temperature, (3.4), as are close to the conditions of experiments with a cold ball immersed in hot water, described in Chap. 6. The ball with a
temperature of 5°C was placed in a cell filled with water at a temperature of 60°C.

A model of the radial dependence of temperature having the following parameters was analyzed:

$$R = 20.5 \text{ mm}, \quad T_0 = 60°\text{C}, \quad \Delta T = 55°\text{C}, \quad \Delta R = 0, \quad a = 1 \text{ mm}.$$

A specific feature of cold-layer refractograms is the inversion of the laser plane and formation of a loop, which is directly associated with the presence of a caustic. It is well known [1] that the caustic in the case of spherical inhomogeneities coincides with the z-axis and can be visualized experimentally. The inversion of the laser plane on the passage of its rays through the caustic is illustrated by 3D refractograms in the following figures. The loop in Fig. 5.7 is formed by the rays deflected

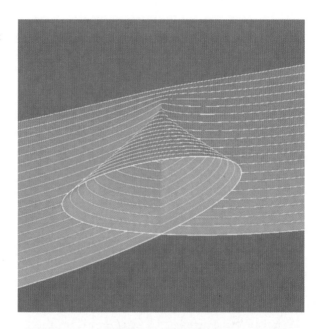

Fig. 5.7 Typical refractogram of a "cold" transparent inhomogeneity of spherical shape

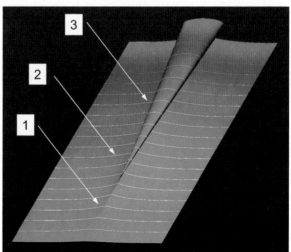

Fig. 5.8 Formation of fold *1*, "beak" *2*, and loop *3* in the refractogram of a "cold" spherical inhomogeneity

toward the center of the inhomogeneity (toward the highest refractive index values) and crossing the z-axis.

The formation of a loop is traced in detail in Fig. 5.8. First, as the distance from the inhomogeneity increases, fold *1* is formed as a result of deflection of the rays toward the center of the inhomogeneity; when the rays intersect the z-axis, a singular point—"beak" *2* appears, and then loop *3* is formed.

Figure 5.9 presents 2D refractograms for the above model parameters and different distances z to the screen: $z = 425$ mm, $z = 525$ mm, $z = 585$ mm, $z = 640$ mm, $z = 725$ mm, and $z = 850$ mm.

The presentation of refractogram sections located close to the caustic in the figure (the plane of observation is normal to the z-axis) makes it possible to trace in

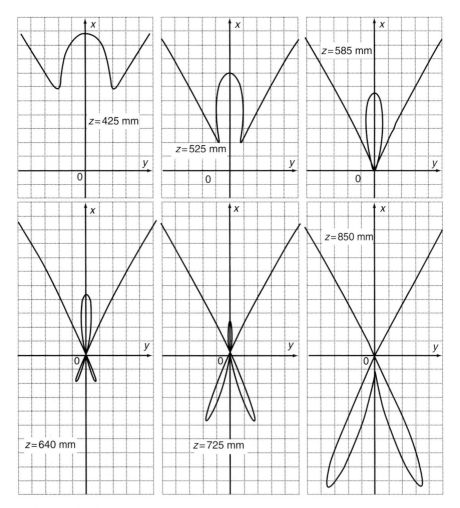

Fig. 5.9 Formation of a refractogram of a laser plane in its caustic region for a "cold" spherical boundary layer. The mesh width in the plots is 1 mm

detail the formation of the loop. The mesh width is 1 mm. A set of 2D refractograms obtained experimentally allows one to reconstruct the respective 3D image and determine the type of the inhomogeneity in hand. The pertinent 3D refractograms are shown in Figs. 5.10 and 5.11.

Fig. 5.10 Formation of extrema in the refractogram loop of a "cold" spherical layer

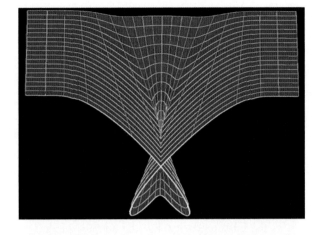

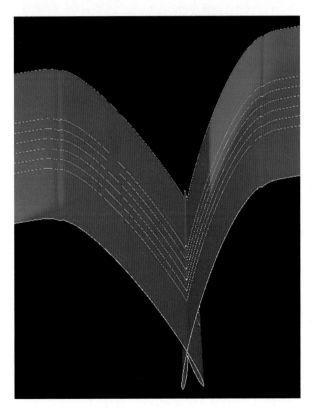

Fig. 5.11 Loop bifurcation in a refractogram of a "cold" spherical layer

If there is an inhomogeneity region wherein the temperature gradient is relatively small or equal to zero, local extrema appear in the refractogram due to the reduction of the laser beam deflection in low-gradient regions (Fig. 5.10). The degree of expressiveness of the extrema varies as a function of the distance from the inhomogeneity, and in the neighborhood of the singular points the loop can bifurcate (Fig. 5.11).

5.1.4 Refractograms of a Transparent Spherical Inhomogeneity with a Temperature Gradient

Let the refractive index profile be defined by the expression

$$n(r) = n_0 + \Delta n e^{-\frac{r^2}{a^2}}, \tag{5.10}$$

where n_0 is the refractive index of the homogeneous medium, Δn is the deviation from the value of n_0 at the center of the inhomogeneity, and a is the characteristic size of the inhomogeneity. For the adopted model, positive Δn corresponds to a situation where the refractive index decreases from the center of the inhomogeneity toward its periphery. In that case, ray trajectories deflect toward the center of the inhomogeneity, and the value of the radial coordinate of the turning point, r_t, is smaller than that of the impact parameter ρ. At sufficiently great distances z_1 the rays can intersect the z-axis, which leads to the inversion of the laser plane and formation of a caustic coincident with the z-axis. The model considered corresponds to thermal lenses in water at a temperature growing higher from the center of the inhomogeneity toward its periphery.

In a transparent spherical inhomogeneity, the shifting of the laser plane, specified by the quantity x_0, causes its projections shown in Fig. 5.12 to change. In this figure, the model parameters are $n_0 = 1.33$, $\Delta n = 0.08$, $a = 7$ mm, and $z_1 = 150$ mm, and x_0 varies by 2-mm steps from 1 to 11 mm (curves 1–6, respectively). Obviously the higher the average refractive index gradient in the propagation region of the laser plane, the greater are its distortions. As follows from expression (5.10), the maximum gradient occurs at

$$r_m = \frac{a}{\sqrt{2}} \approx 5 \text{ mm}. \tag{5.11}$$

At the same value of x_0 (curve 3 in Fig. 5.12), the laser plane is observed to suffer the greatest distortion expressed as its self-crossing to form a loop in the region of negative x values. This effect is explained by the above-described deflection of the laser plane rays toward the center of the inhomogeneity and their crossing of the z-axis. Because of the spherical symmetry of the problem, this deflection comprises components along both the x-axis (maximal for the central rays) and the y-axis (maximal for the peripheral rays).

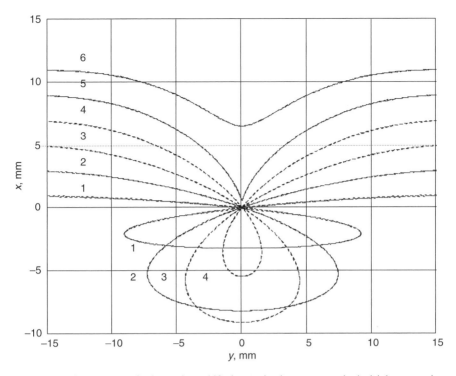

Fig. 5.12 Refractograms of a laser plane shifted stepwise in a strong spherical inhomogeneity with a negative refractive index gradient. $n_0 = 1.33$, $\Delta n = 0.08$, $a = 7$ mm, $z_1 = 150$ mm; x_0 varies by 2-mm steps from 1 to 11 mm (curves *1–6*, respectively)

When the laser plane is moved away from the origin of coordinates (x_0 is increased), its peripheral rays travel a shorter distance in the region of substantial inhomogeneity and, accordingly, undergo a weaker deflection along the *y*-axis. When the laser plane is shifted the center of the inhomogeneity (x_0 is reduced), its loop is observed to flatten (curves *1* and *2*) and conversely, when it is shifted away from the center, the loop extends (curve *4*). As x_0 increases, the loop in the projection degenerates into a "beak" (curve *5*), and as x_0 is further increased, it turns into a smooth curve (curve *6*).

In Fig. 5.13, the model parameters are $n_0 = 1.33$, $\Delta n = -0.01$, $a = 7$ mm, $z_1 = 1.000$ mm, and $x_0 = 1$ mm (curve *1*) and thereupon varies by 3-mm steps from 2 to 14 mm (curves *2–6*, respectively). For the model adopted, negative Δn corresponds to a situation where the refractive index grows higher from the center of the inhomogeneity toward its periphery. In that case, ray trajectories deflect away from the center of the inhomogeneity, and the value of the radial coordinate of the turning point, r_t, is greater than that of the impact parameter ρ. The ray bundle about the *z*-axis is divergent, and so there is no laser plane inversion. This model corresponds to thermal lenses in water at a temperature decreasing from the center of the inhomogeneity

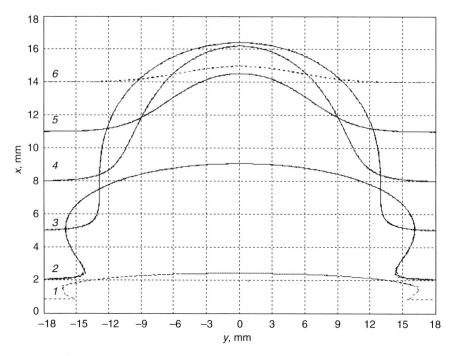

Fig. 5.13 Variation of the projection of a laser plane during its shifting in a weak spherical inhomogeneity with a positive refractive index gradient. $n_0 = 1.33$, $\Delta n = 0.01$, $a = 7$ mm, $z_1 = 1.000$ mm, and $x_0 = 1$ mm (curve *1*) and thereupon varies by 3-mm steps from 2 to 14 mm (curves *2–6*)

toward its periphery. The parameters of the model considered describe a thermal lens with a temperature difference of a few tens of degrees centigrade.

Comparing between the results of laser plane refraction in transparent radial inhomogeneities makes it possible to reveal qualitative differences in shape between its projections for different parameters of the model used. Thus, analyzing a set of projections obtained through measurements based on the method of refractography allows the sense of the inhomogeneity gradient and the order of its magnitude to be determined directly. The method based on the computer processing of experimental curves (Chap. 7) and their comparison with theoretical ones (Chap. 8) enables one to recover the quantitative parameters of the inhomogeneity under study.

5.1.5 Nonmonotonic Cylindrical and Spherical Inhomogeneities

A specific feature of such inhomogeneities is the nonmonotonic character of the radial dependence of their refractive index, $n(r)$, which corresponds to layered structures, possibly having a core, or none. To analyze the general physical laws governing the refraction of laser planes in such media, we use the following inhomogeneity model:

$$n(r) = n_0 + \Delta n e^{-\frac{r^2}{a^2}} + \Delta n_1 e^{-\frac{(r-\Delta R_1)^2}{a_1^2}}, \tag{5.12}$$

where $n_0 = n(\infty)$ and the parameters a and Δn determine the characteristic size of the inhomogeneity core and the maximum deviation of the refractive index in the core from n_0, respectively. Similarly, the parameters a_1 and Δn_1 determine the characteristic size of the first layer and the maximum deviation of the refractive index in this layer from n_0. The parameter ΔR_1 specifies the position of the center of the layer, it being assumed that

$$(a + a_1) < \Delta R_1; \tag{5.13}$$

that is, the core and the layer are taken to be practically "nonintersecting" and capable of being roughly treated as independent structures.

Figure 5.14 presents refractograms of the inhomogeneity at a distance of $z = 150$ mm for the following model parameters: $n_0 = 1.33$, $\Delta n = 0$, $\Delta n_1 = -0.01$, $\Delta R_1 = 8$ mm, and $a_1 = 2$ mm; i.e., this model corresponds to the case where the core is absent, there being but a single nonmonotonic layer with a local refractive index minimum. A physical interpretation of this model may be provided by a situation where, for example, a transparent liquid contains a hot spherical or cylindrical layer whose structure is conditioned by the specific features of the source (a ring-type optical or acoustic emitter). Refractogram (projection) *1* corresponds to the case where the laser plane passes at a distance of 13 mm from the center of the layer. Thereupon the laser plane is shifted by 2-mm steps toward the center up to a distance of 3 mm from it, inclusive. As follows from the given model parameters, the layer can roughly be considered bounded by 6- and 10-mm radii. The laser plane corresponding to projection *1* passes outside of the layer, and so it suffers practically no distortion. Projections *2* and *3* correspond to the situation where the laser plane passes inside

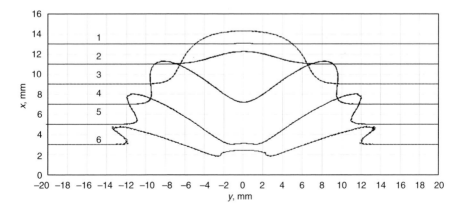

Fig. 5.14 Refractograms of a spherical layer with a local refractive index minimum. Projection *1*—laser plane 13 mm distant from the center of the inhomogeneity, projection *2*—11 mm, projection *3*—9 mm, projection *4*—7 mm, projection *5*—5 mm, projection *6*—3 mm

the layer in a region where the refractive index gradient is negative, its magnitude being greater for projection *3*. In that case, the laser plane rays are deflected away from the center of the inhomogeneity, so that the projections are similar to those considered above for thermal boundary layers. Projection *4* corresponds to the situation where the refractive index gradient in the region of importance for refraction is positive and the rays are deflected toward the center of the inhomogeneity. This projection is similar to the ones considered above for transparent thermal lenses in water. Projections *5* and *6* are specific to nonmonotonic layers in particular and correspond to the situation where the laser plane crosses the spherical layer under study two times. When the laser plane passes through the center of the inhomogeneity ($x = 0$), the projection degenerates into a straight line, which is obvious from symmetry considerations.

Figure 5.15 shows refractograms of an inhomogeneity at a distance of $z_1 = 150$ mm for the model parameters $n_0 = 1.33$, $\Delta n = -0.02$, $a = 3$ mm, $\Delta n_1 = -0.01$, $\Delta R_1 = 8$ mm, and $a_1 = 2$ mm corresponding to a model with a cold core and a single cold layer (two local refractive index maxima).

Projection *1* corresponds to the case where the laser plane passes at a distance of 12 mm from the center of the inhomogeneity. Thereupon the laser plane is shifted by 2-mm steps toward the center of the inhomogeneity up to a distance of 2 mm from the center, inclusive.

Projections *1* and *2* practically do not differ from the corresponding ones in Fig. 5.12, but starting with projection *3*, the presence of the core becomes manifest. Characteristic loops develop in projection *6* as a result of the laser plane inversion described by Evtikhieva [2, 3].

The results obtained can be used for the analysis of multilayer inhomogeneities. In that case, individual refractograms can be constructed for each layer subject to

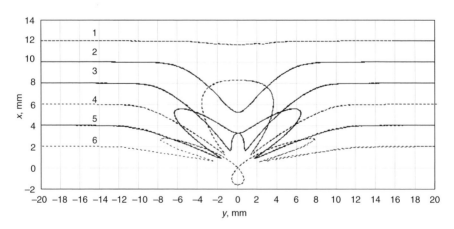

Fig. 5.15 Refractograms of an inhomogeneity with two local refractive index maxima (core and layer). Projection *1*—laser plane at a distance of 12 mm from the center of the inhomogeneity, projection *2*—10 mm, projection *3*—8 mm, projection *4*—6 mm, projection *5*—4 mm, projection *6*—2 mm

condition (5.13), and the full "portrait" of the inhomogeneity will then bear the characteristic features of its structural components. Thus, to analyze complex layered inhomogeneities, use can be made of a set of elementary refractograms [2–4]. The refractograms obtained are a "portrait" in a way that characterizes the specific features of the refractive index profile.

5.2 Refractograms Based on Cylindrical SLR

The refraction of cylindrical SLR in spherically layered inhomogeneities is described on the basis of relations (5.3)–(5.8), subject to the condition that each refracting ray is associated with a parameter characterizing its position on the cylindrical SLR surface. By analogy with what has been done when considering laser plane refraction (Sect. 5.1.1), we will use for this parameter the polar angle φ reckoned counterclockwise from the positive direction of the x-axis. The main geometrical parameters of the problem are shown in Fig. 5.16.

The center of the ball of radius R, around which exists a boundary layer, coincides with the origin of coordinates. The equation of the cylindrical surface formed by the SLR rays parallel to the z-axis is given by

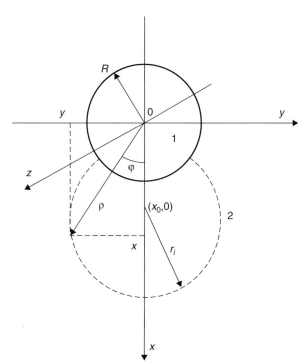

Fig. 5.16 Main geometrical parameters for calculating the refraction of cylindrical SLR: *1*—ball, *2*—cylindrical laser beam

$$(x - x_0)^2 + y^2 = r_i^2, \tag{5.14}$$

where r_i is the radius of the ith SLR cylinder whose axis with the coordinates $(x_0, 0)$ coincides with the central SLR ray.

The current coordinates on the cylindrical surface are

$$\begin{aligned} x &= \rho \cos \varphi, \\ y &= \rho \sin \varphi \end{aligned} \tag{5.15}$$

where ρ is the ray impact parameter, which is actually equal to the polar radius.

It follows from expressions (5.14) and (5.15) that the ray impact parameter ρ on the given cylindrical surface depends on the angle φ as follows:

$$\rho_{1,2}(\varphi) = x_0 \cos \varphi \pm \sqrt{r_i^2 - x_0^2 \sin^2 \varphi}; \tag{5.16}$$

that is, generally two branches of the function $\rho(\varphi)$ exist.

Since it is evident that $\rho_{1,2}(\varphi) > 0$, then, if $\cos \varphi > 0$, which corresponds to the situation shown in Fig. 5.17, the function $\rho(\varphi)$ can have two branches, provided that

$$x_0 \cos \varphi > \sqrt{r_i^2 - x_0^2 \sin^2 \varphi}. \tag{5.17}$$

Inequality (5.17) will hold true if the minimal radius of the SLR cylinder satisfies the condition

$$r_{i\,min} < x_0 - R. \tag{5.18}$$

Actually the principal condition for the existence of a second branch of (5.16) is

$$x_0 > R; \tag{5.19}$$

that is, in other words, the central ray (axis) of the cylindrical SLR, whose coordinates are $(x_0, 0)$, should not be shut out by the ball. In that case, a single value of the ray parameter φ is associated with two values of the impact parameter $\rho(\varphi)$, which gives rise to "loops" in refractograms (Fig. 5.17).

Figure 5.18 presents refractograms of a thermal boundary layer at a hot ball in water for different model parameters of the temperature distribution in the layer in the case where the central SLR ray is shut out by the ball. In this case, the function $\rho(\varphi)$ is single-valued, and so there are no loops in the refractograms.

The two-valuedness of the function $\rho(\varphi)$ is undesirable in the solution of the inverse problem (recovery of layer parameters from refractograms), and therefore the situation where $x_0 < R$, i.e., the central SLR ray is shut out by the sphere, is preferable in practical measurements. In that case, the function $\rho(\varphi)$ is single-valued:

$$\rho(\varphi) = x_0 \cos \varphi + \sqrt{r_i^2 - x_0^2 \sin^2 \varphi}. \tag{5.20}$$

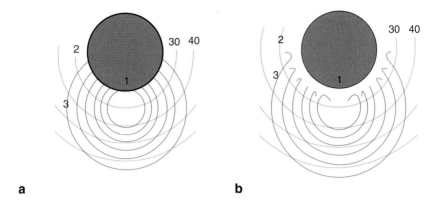

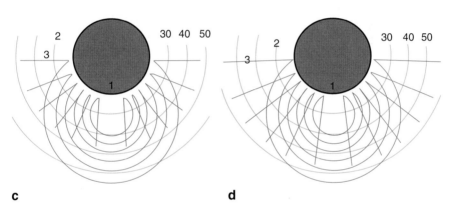

Fig. 5.17 Theoretical refractograms of cylindrical SLR in the case where the central ray is not shut out by the ball for different positions of the ball: *1*—ball, *2*—grid lines, *3*—SLR projections on the screen (2D refractograms). **a** Unperturbed cylindrical SLR. **b** $T = 90°C$, $R = 20$ mm, $a = 2$ mm, $z_1 = 100$ mm. **c** $T = 90°C$, $R = 20$ mm, $a = 0.5$ mm, $z_1 = 200$ mm. **d** $T = 90°C$, $R = 20$ mm, $a = 0.2$ mm, $z_1 = 200$ mm

5.3 Refractograms of Linear Multiple-Point SLR

The numerical modeling of refractograms based on multiple-point SLR (linearly structured laser radiation) is performed for radially inhomogeneous media, with due regard for the possible weak azimuthal inhomogeneity. Examined are transparent cylindrically or spherically symmetric inhomogeneities in air and liquids that can correspond, for example, to hot or cold air flows and temperature inhomogeneities in water. Modeled in addition are radial inhomogeneitites in the form of thin boundary layers near hot or cold opaque bodies in liquids and air.

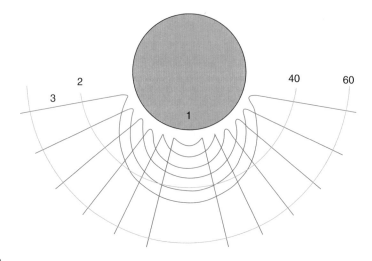

a

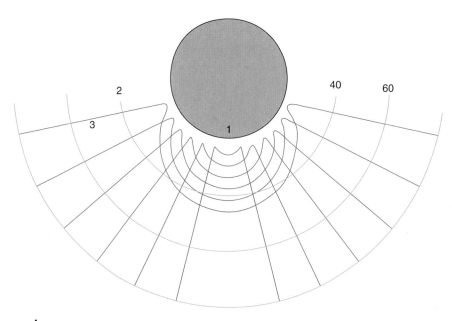

b

Fig. 5.18 Theoretical refractograms of cylindrical SLR in the case where the central ray is shut out by the ball: *1*—ball, *2*—grid lines, *3*—SLR projections on the screen (2D refractograms). **a** $T = 90°C$, $R = 20$ mm, $a = 0.5$ mm, $z_1 = 200$ mm. **b** $T = 90°C$, $R = 20$ mm, $a = 0.2$ mm, $z_1 = 200$ mm

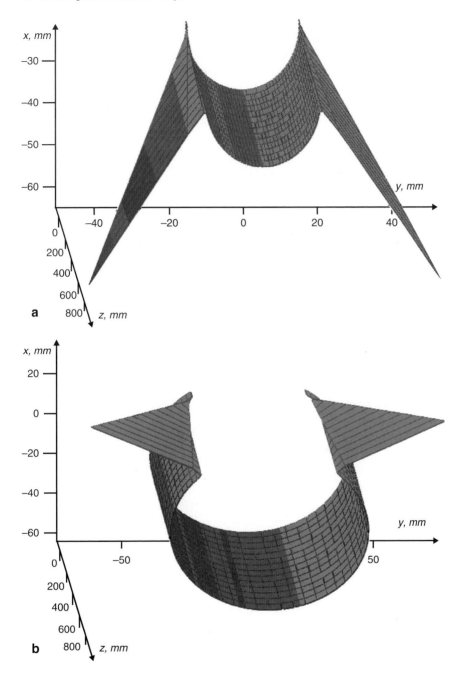

Fig. 5.19 Theoretical 3D refractograms of cylindrical SLR at $T = 90°C$, $R = 20$ mm, and $a = 0.5$ mm. **a** $r_i < x_0 < R$. **b** $x_0 < R < r_i$

It is assumed that the center of the inhomogeneity coincides with the origin of coordinates, the screen where on refractograms are observed is located at a distance of z from the center of the inhomogeneity, the coordinates of the screen center being $(0, 0, z)$. When linear SLR passes through a transparent inhomogeneity, the SLR beam with the parameter r_n equal to the radial coordinate of the SLR source is deflected by an amount of Δr_n given by

$$\Delta r_n = z\tan\gamma_n, \qquad (5.21)$$

where γ_n is the deflection angle of the beam in the inhomogeneity, which can be found on the basis of the methods described in Sect. 4.1.2 to be

$$\gamma(r_n) = \int_{r_t}^{\infty} \frac{2n_0 r_n \dfrac{dn}{dr} dr}{n(r)\sqrt{n^2(r)\,r^2 - n_0^2 r_n^2}}. \qquad (5.22)$$

Figure 5.20 presents a theoretical refractogram for a model with an exponential radial and additional azimuthal inhomogeneity, the latter being manifest in that the spatial scale a of the inhomogeneity (or the thickness of the boundary layer) varies, depending on the angle φ:

$$a = a(\varphi) = a_0 + b\sin\varphi, \qquad (5.23)$$

where $a_0 > 0$, $b > 0$, or $b < 0$.

The physical cause of the azimuthal variation of the characteristic size of the inhomogeneity, or boundary layer thickness, can be, for example, free convection of the liquid in the gravitational field.

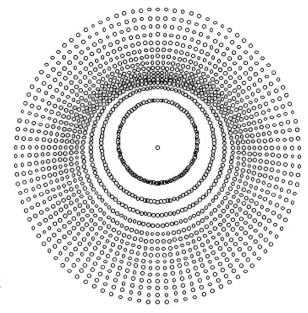

Fig. 5.20 Refractogram of multiple-point SLR for a spherical transparent inhomogeneity in water with an additional azimuthal inhomogeneity: $a = 5 + 2\sin\varphi$

Fig. 5.21 Refractogram of a boundary layer at a hot ball in air, visualized with square-grid multiple-point SLR (mesh width is 1 cm)

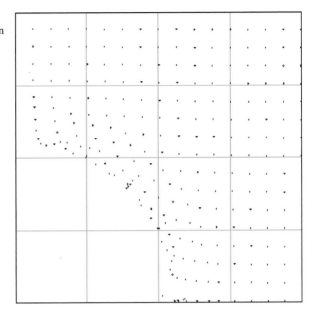

The applicability condition of the calculation algorithm based on relations (5.21)–(5.23) is the requirement that the azimuthal gradient should be small in comparison with its radial counterpart.

Figure 5.21 illustrates the possibilities of using the laser refractography technique to study thermal processes occurring in air. Inasmuch as the temperature dependence of the refractive index in air is weaker by three orders of magnitude than that in water, to observe visually perceptible effects requires temperatures on the order of a few hundreds of degrees centigrade and a few meters distant screens.

Modeled in Fig. 5.21 is a refractogram of a boundary layer near an opaque ball in air, visualized with a square-grid multiple-point SLR. The characteristic size of the inhomogeneity is $a = 6.8$ mm, the ball surface temperature, 300°C, air temperature, 20°C, screen mesh width, 5 mm, and the distance to the screen, 2 m.

The surface of the ball is not marked in any special way, but it is distinctly visualized owing to the displacement of dots in the boundary layer. It follows from this figure that boundary layers at spherical and cylindrical surfaces should better be visualized with square-grid SLR, for in this case, thanks to the uneven displacement of dots, distinct areas of their concentration and rarefaction appear. At the same time, quantitative estimates of radial temperature gradients are more convenient to make using concentric-grid SLR that allow determining the displacement of dots located at a specified distance from the center of the inhomogeneity.

The numerical modeling of linear SLR refractograms, using as an example temperature inhomogeneities in water and air, can help one to draw a conclusion as to the possibility of visualizing transparent inhomogeneities and boundary layers in water in the temperature range from 15 to 100°C on screens a few tens of centimeters distant from the inhomogeneity and in air at temperatures of a few hundreds of degrees centigrade on screens located at distances from a few tens of centimeters to a few meters from the inhomogeneity.

To quantitatively estimate gradients in the refractive index field, use should be made of a numerical analysis capable of determining the displacement of beams from the data of the digital processing of refractograms. Numerical modeling makes it possible to draw a conclusion as to the possibility of visualizing not only radial, but also azimuthal refractive index gradients [5–7]. If the azimuthal gradient is sufficiently small in comparison with the radial one, the beam displacement can be numerically computed using as a basis data obtained with concentric-grid SLR. This approach enables one to model boundary layers around spherical and cylindrical bodies in conditions of free convection in the gravitational field.

References

1. Y. A. Kravtsov and Y. P. Orlov, *Geometrical Optics of Inhomogeneous Media* (Nauka, Moscow, 1980) [in Russian].
2. O. A. Evtikhieva, "Refraction of laser plane in a spherically inhomogeneous thermal boundary layer," *Measurement Techniques*, **49**(5), pp. 466–471 (2006).
3. O. A. Evtikhieva, "Modeling of the Refraction of a Laser Sheet in Transparent Radially Non-uniform Media," *Measurement Techniques*, **49**(10), pp. 1027–1032 (2006).
4. O. A. Evtikhieva, "Diagnostics Specificities of Layered Spherical and Angular Inhomogeneities," *Measurement Techniques*, **49**(12), (2006).
5. O. A. Evtikhieva, B. S. Rinkevichyus, and V. A. Tolkachev, "Visualization of Nonstationary Free Convection in Liquids with a Structured Laser Beam," in CD Rom *Proceedings of ISFV-12, Goettingen, Germany, 2006*. Paper No. 51.
6. I. L. Raskovskaya, B. S. Rinkevichyus, A. V. Tolkachev, and E. S. Shirinskaya, "Refraction of a cylindrical laser beam in a boundary temperature layer," *Optics and Spectroscopy*, **106**(6), pp. 904–909 (2009).
7. O. A. Evtikhieva, B. S. Rinkevichyus, and V. A. Tolkachev, "Visualization of Nonstationary Convection in Liquids near Hot Bodies with Structured Laser Radiation," *Vestnik MEI*, No. 1, 65–67 (2007).

Chapter 6
Laser Refractographic Systems

6.1 Structural Elements of Laser Refractographic Systems

6.1.1 General Construction Principles of Refractographic Systems

The laser refractographic method for the diagnostics and visualization of optically inhomogeneous transparent media under study [1] is implemented through the following operations:

- Probing of the medium with structured laser radiation, which in a most simple case is a narrow axially symmetric laser beam, an astigmatic laser beam (a laser plane), or a conical or cross-shaped laser beam
- Recording of the SLR refractogram on a ground glass screen with a digital photo or video camera
- Processing of the refractogram obtained with special software

The measuring system intended for investigations into optical inhomogeneities by the method of laser refractography should contain the following main components: a laser radiation source, an optical SLR-forming unit, and an SLR positioning system, a diffuse scattering screen, a digital photo or video camera, a computer, and special software. The object of study is placed between the SLR-forming optical unit and the ground glass screen.

The SLR refractogram can be visually observed on a reflecting or transmitting ground glass screen and recorded with photographic equipment. If the size of the SLR beam is greater than that of the optical inhomogeneity under study, there will be observed refractogram sections on the screen corresponding to the optically homogeneous regions of the medium, wherein SLR suffers no distortion. These sections are later used to determine the magnitude of variation of the other refractogram sections corresponding to the SLR rays passing directly through the optical inhomogeneity.

B. S. Rinkevichyus et al. (eds.), *Laser Refractography,*
DOI 10.1007/978-1-4419-7397-9_6, © Springer Science+Business Media, LLC 2010

6.1.2 Radiation Sources

The main requirements imposed upon the radiation sources used in refractographic systems are that radiation should have narrow directivity, i.e., high spatial coherence [2], and that the axis of the directivity diagram should be highly stable. These requirements are fulfilled by the majority of laser sources operating in a single-mode regime, i.e., generating a single low-order transverse mode designated as TEM_{00} [3]. The properties of such a laser beam have been considered in Sect. 1.2. The number of longitudinal modes has practically no effect on refractographic measurements.

Table 6.1 lists the main parameters of commercial lasers used in refractography.

Ruby and neodymium lasers operate in pulsed modes and are used to study non-stationary optical inhomogeneities occurring in gaseous media. Frequency-doubled neodymium lasers generating visible radiation (green line) have found wide application in flow visualization systems. Argon lasers have sufficiently high power to allow producing structured laser radiation over a great area, up to a few square meters. Helium–neon lasers possess the best metrological characteristics, especially as regards the stability of the directivity diagram, one of the most important characteristics in refractography.

Especially worthy of note are semiconductor lasers [4] whose radiation quality is being constantly improved. These lasers are small in size (a few tens of millimeters) and consume little electric power. Single-mode semiconductor lasers generating radiation of fixed wavelengths are already commercially available. Nonthermally-stabilized semiconductor lasers can be used to solve a number of refractographic problems.

Detailed technical specifications of the above-indicated lasers can be found on their manufacturers' web sites, some of which are indicated in [5–8].

6.1.3 Optical SLR-Forming and Positioning Units

We will consider the implementation of optical SLR-forming units using as an example the formation of plane-structured laser radiation (laser plane). The theory of optical systems for forming a single laser plane with specified parameters has been considered in Sect. 2.4. Since there are no such systems in serial production,

Table 6.1 Main parameters of commercial lasers

Nos.	Laser type	Wavelength (µm)	Divergence (mrad)	Power (W)	Pulse duration (s)
1	Ruby	0.6943	10	10^5–10^{10}	10^{-3}–10^{-8}
2	Neodymium	1.06; 0.53	10	10^5–10^{10}	10^{-6}–10^{-8}
3	Argon	0.488; 0.5145	5	0.1–10	∞
4	Helium–Neon	0.6328	4	0.0001–0.05	∞
5	Semi-conductor	0.5–0.9	5–60 deg	0.0005–5	∞

they are designed individually to suit the problem in hand [9]. Described below are original optical systems devised for operation with He–Ne and semiconductor lasers. These systems are being used to investigate thermal boundary layers in liquids.

Optical System for Semiconductor Lasers. Such lasers produce highly divergent radiation [4]. To obtain a wide nondivergent laser plane of constant breadth, use should be made of one more long-focus large-aperture lens whose front focus coincides with the back focus of the short-focus lens forming part of the laser module outfit. To this end, a special long-focus plano-convex lens, 117 mm in focal length, was calculated by the method described in Sect. 2.4 and made of Grade K8 glass. The use of this lens in combination with the semiconductor laser module allowed forming a laser plane of constant width equal to 52 mm.

Optical System for Forming Two Laser Planes. Figure 6.1 presents the optical scheme of a unit for forming two laser planes located at a distance of 60 mm from each other, their orientation (vertical or horizontal) being selected to suit the actual experimental conditions. Such laser plane arrangement enables one to easily compare between distorted and undistorted laser planes; i.e., to employ the more accurate differential method of processing refraction patterns.

Radiation source *1* is a semiconductor laser module. Replaceable spherical lens *2* makes it possible to focus radiation at a distance from 0.25 to 1 m and determines the thickness of the laser beam in this zone, which does not exceed 1 mm. Plano-convex cylindrical lens *3*, 15 mm in the radius of curvature of its cylindrical refracting surface, develops the laser beam into a plane. Special prism *4* splits the developed laser beam into two. Two parallel laser planes spaced 60 mm apart are thus formed, their orientation being the same. When use is made of a lens with a focal length of 0.42 m, the focusing area is located at a distance of 270 mm from the outer edge of prism *4*. The opening angle of the planes is 22°.

This angle can be enlarged by using a cylindrical lens with a refracting surface of smaller radius of curvature. To extend the functional capabilities of the laser-plane-

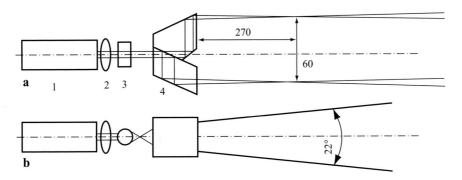

Fig. 6.1 Optical scheme of the unit for forming two laser planes. **a** Side view. **b** Top view. *1*—laser module, *2*—replaceable spherical lens, *3*—plano-convex cylindrical lens, *4*—prism block (parameters in mm)

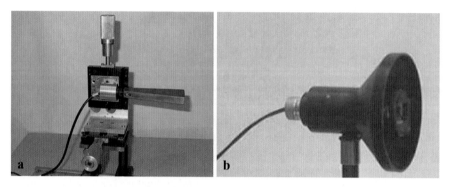

Fig. 6.2 Units for forming **a** a single laser plane and **b** two laser planes

forming unit, provision is made for using replaceable cylindrical lenses differing in the radius of curvature of the refracting surface, such as plano-convex cylindrical lenses 15 and 12 mm in the radius of curvature of the refracting surface. The use of neutral filters between laser module *1* and spherical lens *2* is also admissible.

Figure 6.2a shows the appearance of the unit for forming a single laser plane, complete with the laser module and a heat sink, mounted on a two-coordinate positioner that allows moving the unit, hence the laser plane, in two mutually perpendicular directions (vertical and horizontal), as well as rotating it through any angle about its longitudinal axis. The appearance of the unit for forming two laser planes is shown in Fig. 6.2b.

6.1.4 Digital Refractogram Recording Equipment

As noted earlier, laser plane refractograms are observed on a ground glass screen and recorded with a video camera. The choice of the screen is governed by the experimental conditions. Laser plane refractograms are more convenient to photograph by transmitted light, for in that case their geometrical dimensions suffer no distortion. The distance from the rear wall of the water-filled cell to the screen is selected so as to maximize the size of the refractogram observed on it. While refractogram distortions are small, the screen is moved away from the cell until the image starts smearing out because of the divergence of the laser beam.

The screen is usually equipped with two mutually perpendicular graduated metal scales with a scale-division value of 1 mm. These are necessary for determining the scale of the image in video recording. Where such a screen is used, one can simultaneously carry out the video recording of a multimeter scale.

Refractograms were recorded with various digital equipment, such as the model KONICA Minolta Dimage Z20 photo camera (Fig. 6.3a), SONY DCR-VX 2000 semi-professional video camera (Fig. 6.3b), and Videoscan-285/B-USB black-and-white video camera for scientific research (Fig. 6.3c). The methods of operating

Fig. 6.3 Digital photographic and video equipment used in experiments. **a** Konica Minolta Dimage Z20 photo camera. **b** SONY DCR-VX 2000 video camera. **c** Videoscan-285/B-USB video camera

them are described in Chap. 7. The SONY video camera allows obtaining color refractograms, but from the standpoint of the computer processing of information, it is expedient to use research-oriented black-and-white digital video cameras capable of producing images with the filename extension .bmp or .jpg. For this reason, in some experiments use was made of the Videoscan black-and-white video camera.

6.2 Refractographic Systems with Various Types of SLR

6.2.1 Refractographic Systems Based on Plane-Structured Laser Radiation

The methods for forming plane SLR (laser planes) have been considered in Sect. 6.1.3. At present, the technique of operating experimental systems based on the use of laser planes is the best-mastered one.

Figure 6.4 presents a typical block diagram of a laser refractographic system based on the use of plane-structured laser radiation.

Radiation from visible laser *1* passes through optical system *2* that forms laser plane *3* of specified thickness (in the region of the object of study) and width, or opening angle. Components *1* and *2* are mounted on a two-coordinate positioner intended for moving laser plane *3* in two mutually perpendicular directions by means of micrometer screws.

Laser plane *3* passes through optical inhomogeneity *4* under study. Because of the refraction of the laser plane in the inhomogeneity, its shape gets distorted, which is well visualized on ground glass screen *5* located at some distance from the inhomogeneity. The screen is provided with marks serving to determine the scale of the image in the video recording of the refractogram with digital video equipment *6* connected to personal computer *7* provided with special software.

The computer processing of refractograms can provide information on the refractive index gradient in the region of interest, hence on the distribution therein

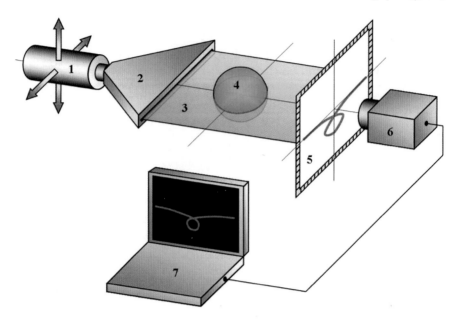

Fig. 6.4 Block diagram of a laser refractographic system: *1*—laser, *2*—SLR-forming unit, *3*—SLR (laser plane), *4*—optical inhomogeneity under study, *5*—ground glass screen, *6*—digital video camera, *7*—personal computer

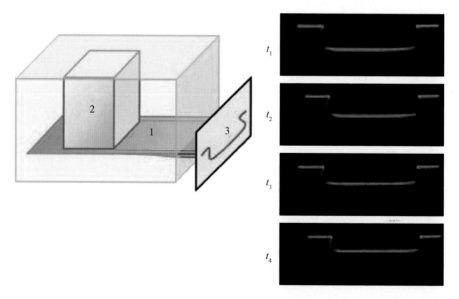

Fig. 6.5 Visualization of an optical inhomogeneity at the bottom surface of a heated parallelepiped in water at different instants of time: *1*—laser plane, *2*—heated body, *3*—screen. $t_1 < t_2 < t_3 < t_4$

of the gradient of a physical field, wherefrom the distribution function of the field itself can be recovered (see Chap. 7).

Figure 6.5 presents as an example the visualization of the boundary layer of water at the bottom surface of a heated metal parallelepiped by means of a laser plane. The refractograms presented refer to different instants of time. One can see from the figure that the laser plane sections passing outside of the boundary layer undergo no distortion.

6.2.2 Refractographic Systems Based on Conical and Cylindrical SLR

Diffraction optical elements can help obtain various types of structured laser radiation that can be adapted to the specificities of the optical inhomogeneity being investigated [10]. Considered below as an example is a refractographic system based on conical SLR.

Figure 6.6 presents a block diagram of an experimental setup intended for the visualization of thermophysical processes occurring in liquids in conditions of free convection near heated bodies by means of conical SLR obtained with the help of

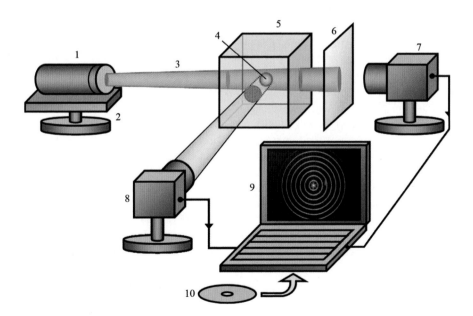

Fig. 6.6 Experimental setup for the visualization of natural convection: *1*—semiconductor laser, *2*—coordinate table, *3*—SLR, *4*—heated body, *5*—transparent water-filled cell, *6*—semitranspar-ent screen, *7* and *8*—digital video cameras, *9*—personal computer, *10*—special software

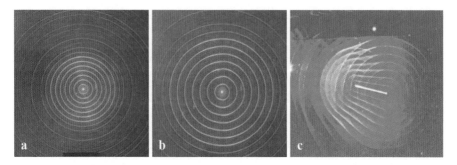

Fig. 6.7 a and **b** Cross-sections of conical laser beams formed at distances of $L=0.5$ m and $L=0.8$ m from the laser module, respectively, and **c** three-dimensional image of conical laser beams visualized by scattered light

optical diffraction elements. The setup consists of semiconductor laser *1*, complete with an optical diffraction element, mounted on coordinate table *2*, SLR *3*, heated body *4*, transparent water-filled cell *5*, semitransparent screen *6*, digital video cameras *7* and *8* whose signals are entered into personal computer *9*, and special software *10*.

Note the small overall dimensions of the laser module. The semiconductor laser, complete with the optical diffraction element, measures 70 mm in length and 20 mm in diameter. It consumes less than 1 W of electric power to generate 20 mW radiation.

Figure 6.7a and b present images of cross-sections of conical laser beams formed with an objective lens 400 mm in focal length and 80 mm in diameter on a ground glass screen located at different distances L from the laser module, and Fig. 6.7c shows the visualization of conical SLR by scattered light.

Observed in the photographs of Fig. 6.7 is the distortion of the cross-sectional shape of conical laser beams at different distances. Undistorted cross-sectional shape is obtained only when the ground glass screen is placed at the focal point of the original laser radiation (at $L=0.5$ m). Consequently, to obtain quantitative information about the cross-sectional shape of conical beams, the original beam should be focused onto the ground glass screen. A similar situation also occurs when using probe radiation in the form of cylindrical laser beams.

Cylindrical laser beams can be obtained from conical beams by placing a spherical collecting lens in their propagation path. By appropriately selecting the distance between the laser module and the spherical lens one can turn conical beams into cylindrical ones.

A specific feature of this system is the presence of another SLR image-recording channel (video camera *8*, Fig. 6.6). This provides additional information about the propagation specificities of SLR in optically inhomogeneous media.

Figure 6.8 illustrates the methods of illuminating the objects under study (a metal ball and spherically bottomed cylinder) mounted on special fixtures in a water-filled cell.

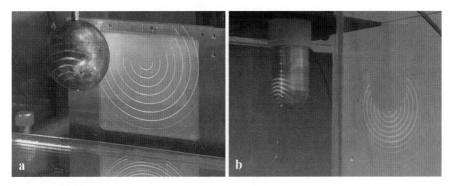

Fig. 6.8 Illumination of various objects with conical and cylindrical laser beams formed with the aid of diffusion optical elements. **a** Metal ball (conical laser beams). **b** Spherically bottomed cylinder (cylindrical laser beams)

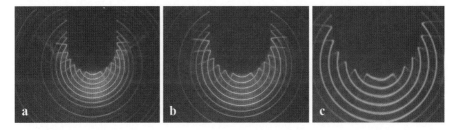

Fig. 6.9 Refractograms of conical SLR near a spherically bottomed cylinder heated to 60°C for different distances L between the laser module and the screen. **a** $L=0.4$ m. **b** $L=0.6$ m. **c** $L=1.0$ m

Figure 6.9 presents refractograms of an optically inhomogeneous boundary layer near a heated (60°C) spherically bottomed cylinder immersed in water at room temperature.

One can see from these refractograms that refraction is normal to the spherical surface; i.e., the boundary layer of liquid near the bottom part of the cylinder is a spherically symmetric inhomogeneous layer.

6.3 Refractographic System for Studying Free Convection in Liquids

6.3.1 Schematic Diagram of the Setup

Figure 6.10 presents a block diagram of the experimental setup intended for investigations into thin boundary layers of liquids near heated bodies [11]. Here *1* is a semiconductor laser module generating visible radiation, *2* is an optical system

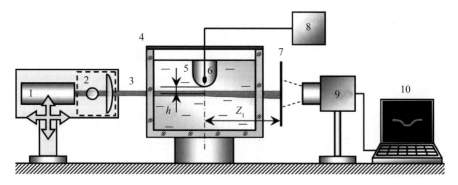

Fig. 6.10 Block diagram of the laser refractographic system: *1*—laser, *2*—laser-plane-forming unit, *3*—laser plane, *4*—water-filled cell, *5*—heated body, *6*—thermocouple, *7*—screen, *8*—digital millivoltmeter, *9*—digital video camera, *10*—personal computer

for forming laser plane *3* with specified thickness (in the neighborhood of the object under study) and width, or laser-plane-opening angle. Components *1* and *2* are mounted on a two-coordinate positioner intended for moving laser plane *3* in two mutually perpendicular directions by means of micrometer screws (for clarity, optical system *2* is shown turned through 90°), glass cell *4* is filled with a transparent liquid (for example, water) and is covered with a lid to which object *5* of study is fixed. The object is heated, and the existence in its neighborhood of an optically inhomogeneous region in water is thus ensured. The temperature of the object is measured with thermocouple *6* connected to meter *8*. The temperature of water is measured with a mercury thermometer with a scale-division value of 0.5°C.

6.3.2 Radiation Sources

Figure 6.11 presents the photograph of the experimental setup. The radiation source used in the below experimental setup is Line Generator semiconductor laser module having the following specifications:

Radiation wavelength (μm)	0.658
Maximum radiation power (mW)	25
Mode of operation	Continuous-wave
Radiation type	Single-mode
Laser-plane-opening angle (degree)	30
Laser plane thickness	Controllable
Operating voltage (DC) (V)	5

The laser module is fed from a stabilized supply source with a rated voltage of no less than 6.3 V. The operating voltage of 5 V is produced by a special stabilizing circuit.

Fig. 6.11 The photograph of the experimental setup for the visualization of free convection

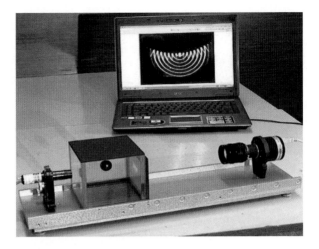

Fig. 6.12 Laser sources. **a** TOPAG semiconductor laser. **b** Laser in a heat sink. **c** LASIRIS™ laser module with a diffraction optical element

The laser module is supplied with two optical attachments. One comprises a cylindrical lens that develops the laser beam into a plane with an opening angle of 30 or 60° and the other is a special attachment with a diffraction optical element that forms two perpendicular laser planes whose opening angle is 24°. Figure 6.12 shows the appearance of laser modules.

6.3.3 Objects of Study and Temperature Monitoring Methods

Free convection was studied in the boundary layers at the surface of heated or cold bodies immersed in a liquid. The objects of study used to demonstrate the measurement method were metal bodies differing in shape (Fig. 6.13).

To take quantitative measurements, it is necessary to monitor the surface temperature of the test objects in whose vicinity convection currents are studied. To this end, a thermocouple was fixed inside the object of study as close as possible to its outer surface, the voltage across the thermocouple terminals being measured with a digital multimeter (millivoltmeter).

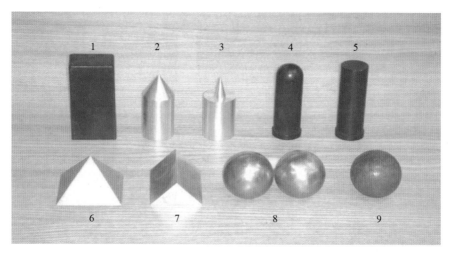

Fig. 6.13 Objects of study: *1*—parallelepiped, *2* and *3*—cone-point cylinders, *4*—spherically bottomed cylinder, *5*—flat-bottomed cylinder, *6*—pyramid, *7*—prism, *8*—two balls, *9*—ball

Special test objects were also designed to accommodate a thermocouple and a heating element. A flat- and a spherically bottomed cylinder were made of Grade Д16АТ material. Their temperature was measured with a digital multimeter complete with a thermocouple intended for measuring temperature in the range from −50 to +240°C.

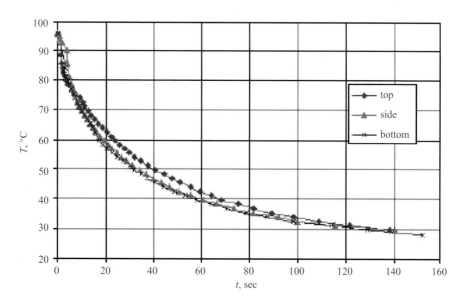

Fig. 6.14 Variation of the temperature of the ball during its cooling

To prolong the cooling of the object of study and prevent the cooling heater from affecting the process, a special cylinder with a spherical bottom of large diameter was designed. It was heated in hot water and then immersed in cold water.

Investigations into free convection were mainly conducted in the neighborhood of a metal ball, with the laser plane passing either below or above it, during the course of its heating and subsequent cooling in cold water, and also during its cooling in hot water.

With the thermocouple calibrated, the cooling time of the ball can be measured with the thermocouple junction placed next to the surface of the ball in its bottom, side, or top part. The relevant graphs are presented in Fig. 6.14. Plotted on the abscissa in this figure is the time of the cooling process and on the ordinate, the ball surface temperature for the three positions of the thermocouple junction.

6.3.4 Refractographic Measurement Procedures

To visualize a nonstationary boundary liquid layer, the video recording of refractograms should be carried out in step with the video recording of the scale of meter *8* reading the time variation of the temperature of object *5* under study. In the course of experiments with heated bodies, the temperature of the bottom surface of the object varied from 80–100° to 30°C, the water temperature in the cell being equal to 20°C.

Consider the technique of refractographic investigations into the boundary layer near a heated steel ball immersed in cold water [10–14]. The positioning of the laser plane in the boundary layer next to the ball at a series of fixed distances from its surface was carried into effect as follows. Use was made of a mask with a hole, whose size and position could be measured, that contacted the lowermost point of the ball surface. By adjusting the position of the laser plane for height with the positioner, the maximum intensity section of the laser plane (its central ray) was made to pass through the hole in the mask. The scale reading of the micrometer screw of the positioner was then fixed and later used to specify the distance from the lowermost point of the ball surface to the center of the laser plane (with respect to its thickness). The boundary layer near the ball was probed with the laser plane passing at different distances h from the lowermost point of the ball, the distance being reckoned from the ball surface: $h = -0.05$ mm—the laser plane is completely shut out by the ball, $h = 0$ mm—the center of the laser plane coincides with the lowermost point of the ball surface, and $h > 0.1$ mm. The initial water temperature in the cell in all the experiments amounted to 23°C. The scale of the multimeter reading the voltage across the thermocouple terminals was synchronously recorded with another video camera.

Figure 6.15 presents exemplary refractograms obtained with the laser plane passing beneath the heated ball cooling in water, as described above. Indicated in the pictures are the time elapsed from the instant the ball was immersed in water and the temperature of the bottom surface of the ball at that instant. The method for

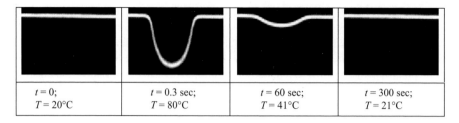

$t = 0$; $T = 20°C$	$t = 0.3$ sec; $T = 80°C$	$t = 60$ sec; $T = 41°C$	$t = 300$ sec; $T = 21°C$

Fig. 6.15 Refractograms of the laser plane passing beneath the heated ball cooling in water for $h = 0.1$ mm at different instants of time

the computer processing of the refractograms presented above are considered in Chap. 8, along with the methods of solving the inverse problems on the recovery of the temperature gradient in the boundary layer.

6.4 Experimental 2- and 3D Refractograms in Boundary Layer Investigations

6.4.1 Gaussian Beam Refractograms in Transmitted and Scattered Light

The interpretation of laser refractograms in actual conditions is not as obvious as in the case of the example considered above. Under the circumstances, it is advisable to investigate the propagation of a narrow laser beam in the inhomogeneity of interest while simultaneously recording the beam images on a screen and in scattered light. To do so when studying liquids, fine particles should be added in a low concentration. For example, polystyrene particles around a micron across are added to water. These particles exert but an insignificant effect on the propagation of the incident beam, but its trajectory in the vicinity of the heated body is plainly visible in scattered light.

The scheme of the experimental setup used provided for the simultaneous recording of the incident laser beam images on a screen and in scattered light. Use was made of a He–Ne laser 1 mW in power and 0.6328 μm in radiation wavelength. The laser beam thickness at the $1/e$ intensity level amounted to 0.3 mm. The laser beam was directed into a water-filled cell. The water temperature was equal to the room temperature $T_r = 20°C$. Lowered into the cell from above was a flat-bottomed metal cylinder 34.5 mm in diameter that could be filled with hot water. The laser beam was made to pass beneath the cylinder at different distances h from its bottom. Because of the temperature difference between the cylinder bottom and the surrounding medium (water), a refractive index gradient developed. The ensuing laser beam deflection observed on the screen was recorded with a video camera. The time variation of the cylinder temperature is shown in Fig. 6.16.

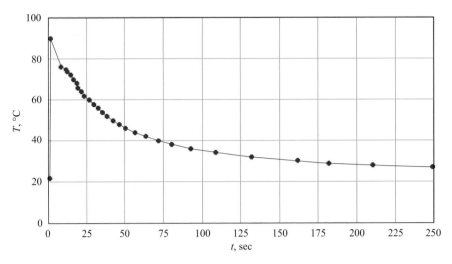

Fig. 6.16 Cylinder temperature as a function of time

t, sec	0	1	8	16	19	26	42	56	71	108	181	249
T, °C	21.5	85	76	70	66	60	50	44	40	34	29	27

Fig. 6.17 Time variations of the laser beam shape and position

The laser beam passed directly beneath the bottom of the cylinder. The results of this experiment are presented in Fig. 6.17.

The beam displacements at various instants of time at the cylinder temperature values indicated are fixed herewith. At the very beginning of the experiment the beam is observed to widen materially because of the sufficiently great temperature gradient and finite beam width. Thereafter, as the cylinder gradually cools and the temperatures level out, the beam slowly recovers its initial condition, following some variations in size and position. The reduction of the cross-sectional size of the beam relative to its initial value at $t = 71$ s points to the existence of a beam-focusing area, owing to the temperature gradient developed.

Figure 6.18 presents scattered-light images of a laser beam passing beneath a heated 50-mm-dia ball at various instants of time that illustrate the beam deflection, focusing, and defocusing effects.

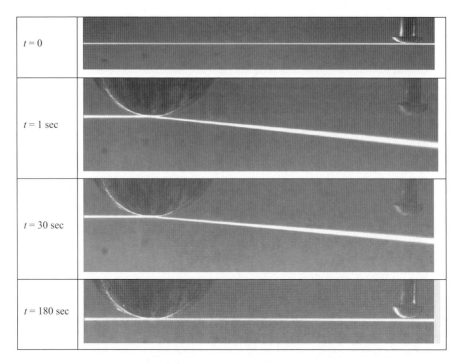

Fig. 6.18 Scattered-light images of a laser beam passing beneath a heated ball at various instants of time

6.4.2   Plane SLR Refractograms in Transmitted and Scattered Light

The block diagram of the experimental setup is presented in Fig. 6.19. A He–Ne laser 1 mW in power and 0.6328 μm in radiation wavelength was used in the experiments. The laser beam thickness at the $1/e$ intensity level was 0.3 mm.

Referring to Fig. 6.19, laser 1 is located at a distance of L_1 from water-filled cell 3. The water temperature in the cell is equal to the room temperature $T_r = 20°C$. The inside length of the cell is L_2. Object 4 under study is lowered into the cell from above. The diameter of the object is $D = 50$ mm. Laser plane 2 is made to pass beneath the object at various distances h from its bottom. The distance from the axis of the object to the inside wall of the cell is L_3.

Screen 5 located at a distance of L_4 from the outside wall of the cell serves to observe deflections of the laser beam projection that are recorded with video camera 6 placed at a distance of L_5 from the screen.

Heated Ball in Cold Water. Of great interest in the theory of natural convection are experimental investigations into the boundary layers near heated bodies immersed in liquids [12, 13]. Presented below is the technique of refractographic studies of the

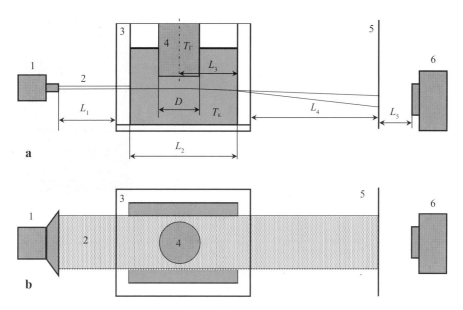

Fig. 6.19 Block diagram of the experimental setup. **a** Side view. **b** Top view. *1*—laser, complete with the laser-plane-forming system, *2*—laser plane, *3*—water-filled cell, *4*—object under study, *5*—screen, *6*—video camera

boundary layer at a heated ball placed in cold water. To this effect, the laser beam is directed to pass over the heated ball as close as possible to its surface. The essence of the experiment is illustrated by Fig. 6.20. It is evident from this figure that the refractograms recorded at different instants of time are of random character.

To get an understanding of the processes leading to such a distortion of the laser plane, an additional experiment was conducted to observe it in scattered light. It

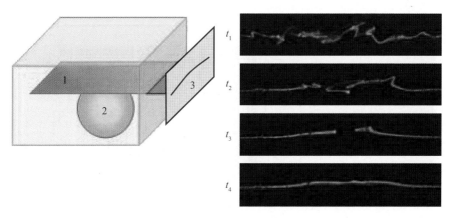

Fig. 6.20 Refractograms of a laser plane passing over a heated ball for different instants of time: $t_1 < t_2 < t_3 < t_4$

Fig. 6.21 Scattered-light images of a laser plane passing over a heated ball immersed in cold water recorded at various instants of time: *1*—laser plane, *2*—heated ball, *3*—digital video camera, *4*—thermics. $t_1 < t_2 < t_3$

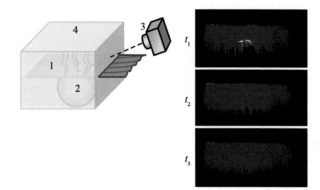

is well known that laser radiation is well scattered by the fine particles present in water (mechanical impurities) or specially added to it. In the given case, to visualize the laser plane, polystyrene latex particles were added to water. Figure 6.21 gives an idea of this experimental technique. The laser plane was video recorded in scattered light from above at an angle of 10° to the horizontal. Clearly evident in the video frames presented is the effect of the so-called thermics (narrow ascending currents in water) on the distortion of the shape of the laser plane.

Cooled Ball in Hot Water. The next experiment illustrates the variation of the refractogram of a laser plane passing at a minimal distance from the top point of the surface of a ball cooled to a temperature of 7°C and immersed in hot water (the water temperature at the beginning of the experiment was 70°C) [1]. The experimental technique is illustrated by Fig. 6.22. The distance from the rear wall of the cell with hot water to the screen amounted to 600 mm. Observed here are regular (reproducible) refractograms whose shape coincides with that of their theoretical counterpart (see Chap. 4). The loop-shaped refractograms formed are characteristic of a cold layer. The self-crossing point corresponds to the position of the caustic.

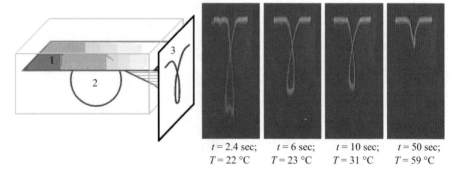

$t = 2.4$ sec; $t = 6$ sec; $t = 10$ sec; $t = 50$ sec;
$T = 22$ °C $T = 23$ °C $T = 31$ °C $T = 59$ °C

Fig. 6.22 Refractograms of a laser plane passing over a cold ball in heated water at various instants of time: *1*—laser plane, *2*—cooled ball, *3*—screen

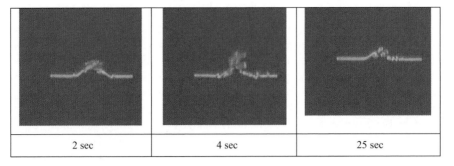

| 2 sec | 4 sec | 25 sec |

Fig. 6.23 Variations of the refractogram of a laser plane passing beneath a cold ball immersed in heated water for different instants of time

A similar technique was also used to study free convection at the bottom surface of a ball cooled to a temperature of 5–7°C and placed in water with a temperature of 70°C. The laser plane here passed at a minimal distance from the bottom point of the ball surface (the laser plane practically touched the ball). Figure 6.23 presents a series of the pertinent refractogram image video frames.

Owing to the cold liquid microcurrents descending from the surface of the ball, the refractograms obtained are also of irregular character. Nevertheless, as in the case of laser plane passing over the ball, the formation of a caustic and a tendency toward the formation of a loop can be traced here as well.

The analysis of the above refractograms shows that what occurs in the given case is an unstable form of natural convection that does not follow from its theoretical treatment (see Chap. 8).

6.5   Refractographic System for Visualization of Liquid Mixing in Twisted Flows

6.5.1   Liquid-Mixing Visualization Procedure

Since the refractive index gradient of turbulent twisted flows in intermixing liquids is a random function, the trajectories and deflection angles of laser beams in such media are random functions as well. In refractographic measurements, subject to visualization and measurement are laser plane deformations observed on a semi-transparent screen and recorded with a digital video camera.

When two liquids differing in refractive index, for example, water and glycerin, are mixed, the solution gradually gets homogenized, its refractive index leveling out in space; i.e., the solution becomes optically homogeneous. The homogenization time of a solution is traditionally determined by measuring its conductivity with a conductometer whose remote probe is placed in the desired zone of the solution. The shortcomings of this method are, firstly, perturbation of the liquid flow because of the finite size of the measuring probe and secondly, the local character of measurements.

The gist of the refractographic method for determining the homogenization time of solutions [15] is that the medium under study is probed with a wide but thin laser plane whose geometrical parameters vary because of the refraction of light by large-scale refractive index inhomogeneities of the medium. At first the refractive index gradient of a mixture of dissimilar liquids (for example, cold and hot water, sweet and salt water, water and glycerin) changes most sharply, and then, as the liquid mixing progresses, the medium becomes optically homogeneous. The time the mixture becomes optically homogeneous characterizes the completion time of the liquid-mixing process.

6.5.2   Schematic Diagram of the Experimental Setup

An experimental setup for the visualization of a twisted liquid flow in a cylinder was for the first time described by Evtikhieva and coworkers [16]. Figure 6.24 presents a block diagram of the modernized setup for studying the process of mixing of two liquids [15].

The laser-plane-forming system consists of gas laser *1* and laser-plane-forming unit *2*. The latter is mounted on coordinate device *3* allowing unit *2* to be moved in three mutually perpendicular directions and turned relative to the twisted flow under investigation.

The mechanical mixing device comprises cylindrical glass vessel *6* filled to two-thirds of its volume with water *7*, or some other liquid, wherein the mixing process is implemented with agitator *9* driven by variable-speed motor *8*. To prevent the laser plane from being distorted upon passage through liquid-filled cylinder *6* with walls of finite thickness, the vessel is placed in rectangular (310 by 310 mm²) tank *4* with flat transparent walls, filled with the same liquid.

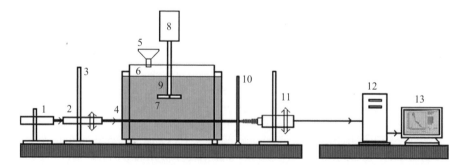

Fig. 6.24 Schematic diagram of the refractographic setup: *1*—laser, *2*—laser-plane-forming unit, *3*—coordinate device, *4*—rectangular vessel, *5*—funnel, *6*—glass vessel, *7*—water, *8*—electric motor, *9*—agitator, *10*—screen, *11*—video camera, *12*—personal computer, *13*—monitor

Placed in the propagation path of the laser plane behind the object of study is ground glass screen *10*. Observed on this screen is a refractogram in the form of a thin strip of light in the case of unperturbed medium, or in the form of a smeared strip of light varying in time because of the refraction of light by refractive index inhomogeneities in the case of perturbed medium. The refractogram is recorded with video camera *11* connected to personal computer *12*. The image of the time-varying refractogram on screen *10* is simultaneously displayed on the screen of monitor *13*. The refractogram is processed with specially developed software.

The experimental procedure involved determining the mixing time of a large quantity of water and a certain dose of saturated solution of common salt by the laser refractometric method and by the traditional method of conductometry, i.e., by measuring the varying conductivity of the mixture, simultaneously [14].

6.5.3 Results of Experimental Studies

Figure 6.25 presents a sample of a series of video frames showing the variation dynamics of the refractogram on screen *10* during the course of mixing of 6 l of water and 20 ml of saturated solution of common salt. The smearing of the refractogram is due to the refraction of radiation by the refractive index fluctuations of the mixture. While the laser plane passes through the homogeneous media (prior to the beginning and after the end of the mixing process), no distortions are observed in the laser plane image. Thus, the duration of the mixing process determines the time for which the laser plane image on the screen stays smeared. The time for which the refractogram has a smeared shape can serve as a quantitative measure of the mixing time.

The analysis of data on the simultaneous measurement of the solution homogenization time by the laser refractometric and the classical conductometric method bears witness to the fact that the laser refractometric method of probing optically inhomogeneous media yields reliable results and can effectively be used in the experimental practice of investigations into mechanical mixing devices [17–19].

6.5.4 Refraction Pattern Processing Algorithm and Program

To determine the mixing time of liquids from experimental results, a special program was worked out, allowing one to filter images, compare between the necessary video frames, and plot refractogram area against mixing time.

Fig. 6.25 Refractogram shape as a function of time. **a** *t*=0. **b** *t*=5.5 sec. **c** *t*=27.8 sec

The input data of the program is a file with experimental refraction video data. When the user points-and-clicks the file with the initial data in the program, the first video frame in the sequence is displayed, after which the user can select the desired type of processing (the time dependence of the laser plane width or area). When the refractogram processing is started, a frame is selected from the video series and a numerical value from 0 to 255 is assigned to each of its pixels. The subsequent operations are performed with video frames whose numerical value exceeds the specified noise level. A pixel with the maximum numerical value is found for each column. Thereafter the numerical values of the rest of the pixels are normalized to this value, and binary quantization is then performed with respect to the $1/e^2$ level. The area of the refractogram is computed in relative units and is equal to the ratio between the number of the positive binarized pixels and the total area of the video frame in pixels.

Figure 6.26 presents the relationship obtained between the refractogram area (in relative units) and the serial number of the video frame. With the frame frequency

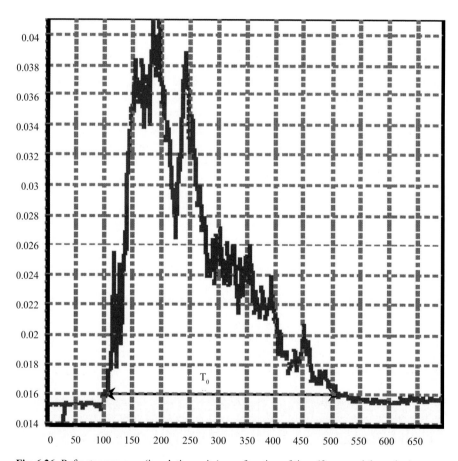

Fig. 6.26 Refractogram area (in relative units) as a function of time (frame serial number)

known, one can plot the time dependence. In the case of fast processes, optimal results are obtained with the noise level set at 1/3 of the maximum possible value, or at 1/2 of the experimental signal maximum. The liquid-mixing time T_0 can be determined from the experimental curve at a level exceeding the initial laser plane area by a specified amount, for example, 10%.

6.5.5 Liquid-Mixing Visualization with Two Crossed Laser Planes

The above examples graphically depict the process of mixing of two liquids in a single section of the flow under study. The method of two crossed laser planes [7] that can help one to trace the variations of two density gradients in two different sections of the flow is more informative. To separate information about the two density gradient components at a single point, use is made of laser planes produced by lasers differing in radiation wavelength. The scheme of the laser illumination of the flow being studied is presented in Fig. 6.27.

The setup used two laser planes produced by two continuous-wave lasers—red He–Ne laser *1* and blue argon laser *2*. The laser plane from the former was directed along the axis of cylinder *3* and that from the latter, at right angles to it. The laser planes passed below agitator *4*.

Typical experimental results are presented in Fig. 6.28. Notice the fact that the laser plane broadening differs between different sections. The vertical laser plane visualizes the refractive index gradients along the horizontal axis, while the vertical one, those along the vertical axis. The use of the laser planes differing in radiation wavelength makes it possible to separate in processing the laser plane broadening into two mutually perpendicular components.

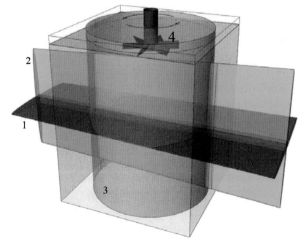

Fig. 6.27 Schematic diagram of two-color laser illumination of a twisted flow: *1*—horizontal laser plane, *2*—vertical laser plane, *3*—cylindrical vessel, *4*—agitator

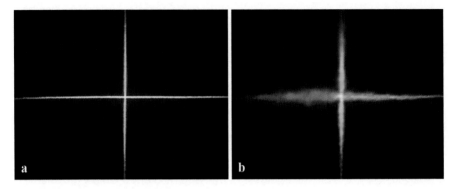

Fig. 6.28 Refractograms of cross-shaped structured laser radiation. **a** Optically homogeneous flow. **b** Optically inhomogeneous turbulent flow

The developed method and program for processing refractograms obtained in digital form with a video camera allows one to determine the time of nonstationarity of the mixing process, caused by the liquid of dissimilar properties introduced into the flow.

6.6 Two-Perspective Laser Refractographic Systems for Monitoring Nonstationary Thermophysical Processes

6.6.1 Two-Perspective System for Obtaining Refractograms of Complex Objects

To investigate thermophysical processes, one-perspective laser refractographic set-ups were previously developed, and the techniques of investigating thermophysical processes taking place in the vicinity of heated bodies immersed in liquids were mastered using this setup, the experimental results being compared with their theoretical counterparts. The results of the experimental studies performed with this setup have been described above.

To further extend the application field of laser refractography requires devising new measuring systems relying for their operation on the tomographic flow diagnostics principle. The possibility of obtaining 3D refractograms elevates the diagnostics of thermal processes in liquids to a new qualitative level.

The refractographic systems considered above are intended for studying simple types of optical inhomogeneity like a single spherical inhomogeneity around a heated ball. In practice, one comes across more complex types of inhomogeneity whose visualization necessitates using modified refractographic systems.

Figure 6.29 shows the optical scheme of a two-perspective system wherein two laser planes are directed at right angles to each other. It consists of two laser mod-

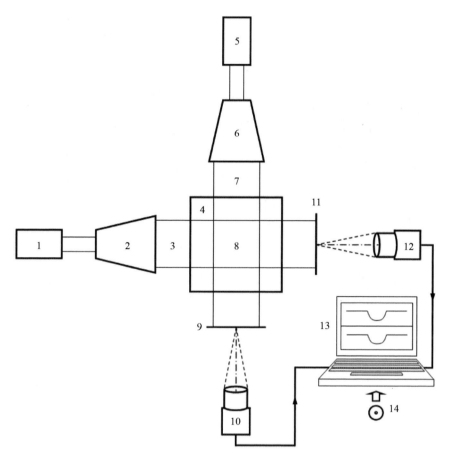

Fig. 6.29 Optical scheme of a two-perspective refractographic system: *1* and *5*—laser modules, *2* and *6*—optical systems, *3* and *7*—laser planes, *4*—water-filled cell, *8*—object under study, *9* and *11*—screens, *10* and *12*—photo cameras, *13*—computer, *14*—special software

ules, *1* and *5*, and two optical systems, *2* and *6*, that form two laser planes, *3* and *7*. Water-filled cell *4* contains object *8* under study. Refractograms are observed on screen *9* and *11* and recorded with digital photo cameras, *10* and *12*, whose signals are entered into computer *13*. Software *14* is specially developed for the simultaneous recording of two refractograms. The appearance of the two-perspective refractographic system is shown in Fig. 6.30.

This system was used to determine the spatial position of heated bodies causing refractive index gradients to develop around them.

Figure 6.31a presents a special test object in the form of a cylinder with three identical small spherically bottomed cylinders fixed to its base. The cylinder made of duralumin was 73 mm in diameter and 37 mm in height. The three small cylinders—projections—on its bottom surface were of steel, their height amounting to

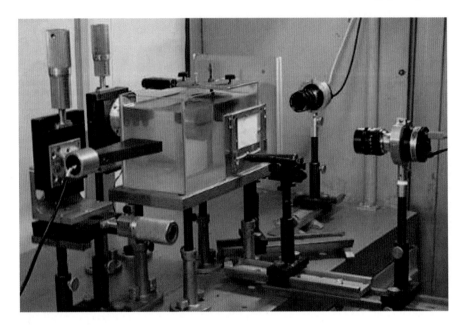

Fig. 6.30 Two-perspective refractographic system

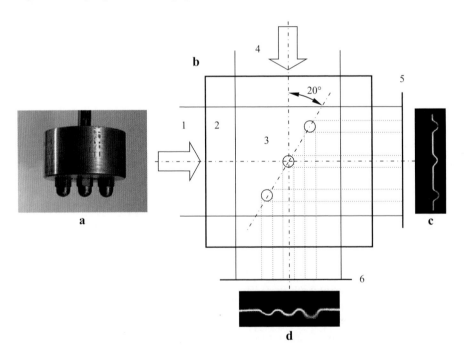

Fig. 6.31 Laser plane refractograms for a cylinder with projections. **a** Photograph of the cylinder with projections. **b** Horizontal projection of the cylinder. **c** Refractogram on screen *9* obtained with camera *10*. **d** Refractogram on screen *11* obtained with camera *12*

15 mm and their spherical tip radius to 6 mm. The arrangement of the projections used to obtain two refractograms is illustrated in Fig. 6.31b.

In the refractogram of Fig. 6.31d, the local deflection of the laser plane for the rightmost projection is substantially greater than for the first two projections on the left, because the former is the farthest from the screen. The refractogram of Fig. 6.31c shows almost equal local laser plane deflections for all the three projections, for the distances from the projections to the screen differ but little. The analysis of the above-presented two-perspective refractograms obtained with two mutually perpendicular probe laser planes passing at the same distance from the surface of the projections on the bottom of the intricately shaped cylinder cooling in cold water makes it possible to determine not only the character of the cylinder cooling process, but also the location of the projections.

6.6.2   Two-Perspective System for Observing Refractograms in Scattered Light

The laser refractography method offers new possibilities for the visualization of optically inhomogeneous flows if light scattering particles are introduced in the flow under study with a view to recording laser radiation scattered by them. In that case, one can observe the propagation of structured laser radiation in the flow and trace all its variations.

To monitor nonstationary ascending currents over a heated metal ball immersed in cold water, an experimental measuring system was devised and made in order to implement the method of the two-perspective probing of the region of interest with two mutually perpendicular laser planes.

Figure 6.32 shows the two-perspective system wherein scattered radiation is recorded with a single digital camera. Here 6 is a glass cell filled with cold water, 5 is a heated metal ball 50.8 mm in diameter, 1 and 3 are optical systems for forming laser planes 2 and 4, 7 is a digital video camera, and 8 and 9 are personal computers. To record the temperature of the heated ball, use is made of millivoltmeter 7 connected to the terminals of a thermocouple fixed inside the ball. The readings of millivoltmeter 7 are recorded with video camera 10. The operation of the video cameras is synchronized by the computer using a special program. Electronic clock 12 serves to record time-dependent processes.

The laser planes are parallel and measure 52 mm in width. To record scattered-light images of the laser planes in water, use is made of a Videoscan digital video camera.

The process under study is characterized by the presence of nonstationary, rapidly changing convection currents in the region above the cooling ball. These are the so-called thermics whose existence does not follow from theoretical calculations concerning natural convection. These thermics were previously visualized with laser planes (see Figs. 6.20 and 6.21).

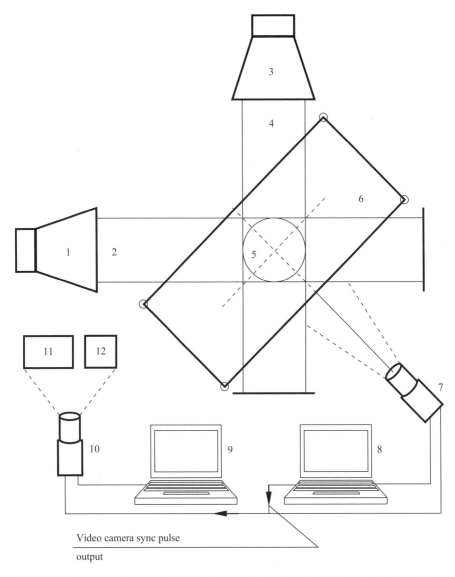

Fig. 6.32 Two-perspective scattered-light refractographic system: *1* and *3*—optical systems, *2* and *4*—laser planes, *5*—heated metal ball, *6*—glass cell filled with cold water, *7*—video camera, *8* and *9*—personal computers, *10*—video camera, *11*—millivoltmeter, *12*—electronic clock

Figure 6.33 presents an exemplary two-perspective refractogram whose shape is determined by the nonstationary character of the thermics over a heated ball cooling in cold water.

A material merit of this system is that it uses but a single digital photo camera to simultaneously record two refractograms. In some cases, however, its shortcoming associated with the narrow directivity of scattered laser radiation becomes manifest, provided that use is made of scattering particles over a few microns across.

Fig. 6.33 Two-perspective laser plane refractogram obtained with a single photo camera for a ball heated to 70°C

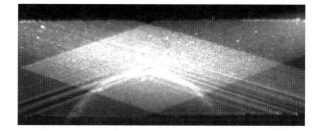

A refractographic system using two digital cameras set at small angles to the optical axis of structured laser radiation proved to be free from this shortcoming.

An essential condition for the recording of the two refractograms was the synchronizing of the operation of the cameras; i.e., making them record two refractogram images occurring at the same instant of time. To this end, use was made of the external sync inputs of the cameras. Special software installed in either computer made it possible to set the frame period shorter than the standard corresponding to a frame rate of 7 frames per second. When selecting the synchronization regime, one should specify the sampling period (five sampling periods available in the given program version determined the frame period) and the sampling pulse duration. Upon completion of the video recording session, the file with the actual frame period could be saved.

A necessary experimental condition is the requirement that the surface temperature of the metal ball should be measured simultaneously with the recoding of the refractograms. In our case, a thermocouple was fixed practically directly under the top surface of the ball. The thermo-emf produced by the thermocouple was

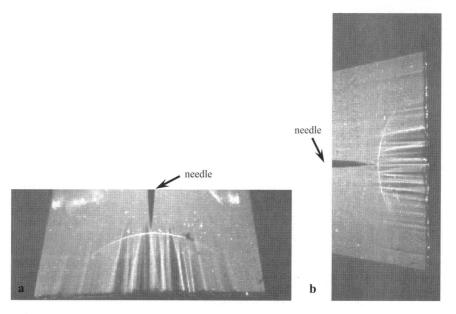

Fig. 6.34 Scattered-light laser plane refractograms for a heated ball. **a** Laser plane *4*. **b** Laser plane *2*

measured with digital millivoltmeter *11* and the current time of the experiment with digital frequency counter *12* operating in the continuous counting mode.

The millivoltmeter and frequency counter readings were recorded with SONY digital camera *10*, the recording being started the instant the sync circuit of cameras *7* and *10* was activated. To subsequently register the refractogram images from either laser plane, a needle was fixed exactly above the top pint of ball *5* at a distance of 1.1 mm from its surface.

The above-described experimental measuring system and the technique of its application in a thermophysical experiment enabled us to obtain shape variation refractograms for two mutually perpendicular laser planes probing the region of nonstationary ascending convection currents (thermics) over a heated ball cooling in cold water. These refractograms are presented in Fig. 6.34.

6.6.3   System for Obtaining 4D Refractograms of Nonstationary Convection Currents in Liquids

The investigation results described in the preceding sections of the book gave grounds for developing a comparatively simple method of 4D diagnostics of nonstationary objects. In essence, the method consists in probing the zone of interest,

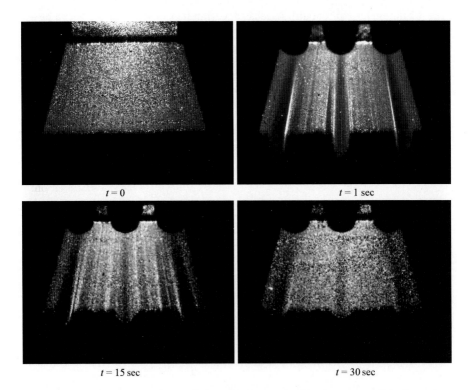

Fig. 6.35 4D refractograms of the cylinder with three projections in scattered light for various instants of time

for example, a convection current, with a laser plane, recording refractograms obtained in scattered light with a digital video camera set at a certain (known) angle to the initial laser plane, and analyzing the refractograms.

Referring to Fig. 6.29, consider a version of constructing a setup for the 4D diagnostics of nonstationary objects. To this end, we exclude from the setup of Fig. 6.29 the measuring channel constituted by laser module *1*, system *2* for forming laser plane *3*, ground glass screen *11*, and video camera *12*. We set video camera *10* at an angle of 30° to the horizontal plane, and raise the radiation power of the semiconductor laser in the system for forming laser plane *7* to 25 mW. We adjust the position of the test cylinder with projections so as to make the line passing through the projection centers be perpendicular to the propagation direction of laser plane *7*. The system for recording refractograms on screen *9* we leave unchanged.

Figure 6.35 presents refractograms of heated body *3* (cylinder with three projections, Fig. 6.31a) for various instants of time. The first refractogram was obtained at a test body temperature equal to the temperature of water in cell *2*. The subsequent 4D refractograms offer a three-dimensional picture of the time variation of the laser plane shape in the course of cooling of heated object *3* in cold water.

6.7 Experimental Refractogram Library

The measuring systems and refractograms described above can help one get an idea of the great capabilities of the method of laser refractographic investigations into optically inhomogeneous media, both stationary and nonstationary. Actually any inhomogeneity can be characterized by its own refractogram depending solely on the type of the inhomogeneity and that of the structured laser radiation used to study it [20, 21]. This makes possible the express diagnosis of the type of optical inhomogeneities, i.e., the type of physical fields giving rise to the given type of optical inhomogeneity. Figure 6.36 presents typical refractograms obtained with various types of structured laser radiation. The heated bodies used for the purpose were immersed in cold water. Use was made of various types of structured laser radiation: a laser plane and cross-shaped and cylindrical SLR. The refraction patterns presented represent typical forms of refractograms.

Refractograms *1–3* in Fig. 6.36 show the variation of the profile of a laser plane passing in the neighborhood of a heated parallelepiped: *1*—under the bottom, *2*—near an edge, *3*—near a vertical side. The role of the edge effects is evident in all the three refractograms. Refractogram *4* corresponds to the deflection of a laser plane passing under a flat-bottomed cylinder. Comparing this refractogram with that for a ball in Fig. 6.15, one can see the substantial difference between them. Refractogram *5* is typical of the refraction of cylindrical SLR by a spherical inhomogeneity, considered in Chap. 5. Refractogram *6* of cross-shaped SLR in an inhomogeneous turbulent flow specific to intermixing liquids with different refractive indices shows the anisotropy of their gradients as a function of coordinates. Refractograms *7–9* are characteristic of cross-shaped SLR wherein the size of the laser beams is greater than the thickness of the boundary layer near the body under study.

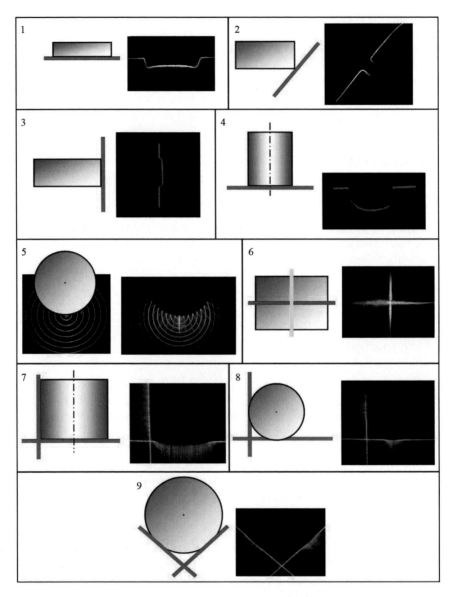

Fig. 6.36 Experimental refractogram library: *1*—horizontal laser plane under the base of a paral-lelepiped, *2*—inclined laser plane near an edge of a parallelepiped, *3*—vertical laser plane along a side of a parallelepiped, *4*—horizontal laser plane under the bottom of a cylinder, *5*—cylindrical SLR near a ball, *6*—cross-shaped SLR in a turbulent flow, *7*—cross-shaped SLR near a vertical cylinder, *8* and *9*—cross-shaped SLR near a ball

References

1. I. L. Raskovskaya, B. S. Rinkevichyus, and A. V. Tolkachev, "Laser Refractography of Optically Inhomogeneous Media," *Kvantovaya Elektronika*, No. 12, 1176–1180 (2007).
2. B. S. Rinkevichius, *Laser Diagnostics in Fluid Mechanics* (Begell House Inc. Publishers, New York, 1998).
3. N. V. Karlov, *Lectures on Quantum Electronics* (Nauka, Moscow, 1995) [in Russian].
4. *Semiconductor Lasers* (Nauka, Moscow, 2005) [in Russian].
5. www.SpectraPhysics.com.
6. www.Plazma.com.
7. www.StockerYale.com.
8. www.Polus.com.
9. O. A. Evtikhieva, A. I. Imshenetsky, B. S. Rinkevichyus *et al.*, "Computer-Laser Refraction Method of Studying Optically Inhomogeneous Flows," *Izmeritelnaya Tekhnika*, No. 6, 15–18 (2004).
10. O. A. Evtikhieva, B. S. Rinkevichyus, and V. A. Tolkachev, "Visualization of Nonstationary Free Convection in Liquids with Structured Laser Beam," in CD-ROM *Proceedings of ISFV-12, Goettingen, 2006*. Paper No. 51.
11. V. I. Artemov, O. A. Evtikhieva, K. M. Lipitsky *et al.*, "Investigation of a Nonstationary Temperature Field in Free Convection Conditions by the Computer-Laser Refraction Method," in *Proceedings of the 8th Scientific-Technical Conference on Optical Methods for Studying Flows*. Ed. by Yu. N. Dubnishchev and B. S. Rinkevichyus (Znak, Moscow, 2005), pp. 478–481 [in Russian].
12. O. A. Evtikhieva, K. M. Lapitsky, and I. L. Raskovskaya, "Laser plane Propagation in the Thermal Field of a Heated Ball in Water," in *Proceedings of the 8th Scientific-Technical Conference on Optical Methods for Studying Flows*. Ed. by Yu. N. Dubnishchev and B. S. Rinkevichus (Znak, Moscow, 2005), pp. 332–335 [in Russian].
13. V. I. Artemyev, G. G. Yankov, O. A. Evtikhieva *et al.*, "Numerical and Experimental Studies of Free Convection in a Liquid near a Heated Cylinder," in *Proceeding of the 4th Russian National Conference on Heat Exchange* (Moscow Power Engineering Institute Press, Moscow, 2006), Vol. 2, pp. 42–46 [in Russian].
14. K. M. Lapitsky, I. L. Raskovskaya, and B. S. Rinkevichyus, "Quantitative Visualization of Transparent Spherical Temperature Layer," in CD-ROM *Proceedings of ISFV-12, Goettingen, 2006*. Paper No. 55.
15. O. A. Evtikhieva, A. I. Imshenetsky, B. S. Rinkevichyus, and A. V. Tolkachev, "Visualization of Mixing in a Twisted Flow by Means of Laser Planes," in CD-ROM *Proceedings of 2nd Russian Conference on Heat Exchange and Hydrodynamics in Twisted Flows, 2005* (Moscow Power Engineering Institute Tech University, Moscow). Paper No. 0320500321.
16. O. A. Evtikhieva, S. V. Orlov, and A. V. Tolkachev, "Investigation of Flows in a Chemical Reactor by the Laser Refraction Method," in *Proceedings of the 6th International Scientific-Technical Conference on Optical Methods for Studying Flows*. Ed. by Yu. N. Dubnishchev and B. S. Rinkevichyus (Moscow Power Engineering Institute Press, Moscow, 2001), pp. 444–447 [in Russian].
17. M. V. Esin and A. V. Tolkachev, "Three-Dimensional Visualization of Nonstationary Flows and Twisted Formations," in *Proceedings of the 6th International Scientific-Technical Conference on Optical Methods for Studying Flows*. Ed. by Yu. N. Dubnishchev and B. S. Rinkevichyus (Moscow Power Engineering Institute Press, Moscow, 2001), pp. 236–239 [in Russian].
18. O. A. Evtikhieva, M. V. Esin, S. V. Orlov, B. S. Rinkevichyus, and A. V. Tolkachev, "Laser Refraction Method for Studying Liquids in Twisted Flows," in *Proceedings of the Third Russian National Conference on Heat Exchange* (Moscow Power Engineering Institute Press, Moscow, 2002), Vol. 1, pp. 197–200 [in Russian].
19. M. V. Yesin, O. A. Evtikhieva, S. V. Orlov, B. S. Rinkevichyus, and A. V. Tolkachev, "Laser Refractometral Method for Visualization of Liquid Mixing in Twisted Flows," in CD-Rom

Proceedings of the 10th International Symposium on Flow Visualization, Kyoto, Aug. 26–29 2002, pp. 1–8. Paper No. F037.

20. O. A. Evtikhieva, B. S. Rinkevichyus, and A. V. Tolkachev, "Visualization of Nonstationary Thermophysical Processes by Computer-Laser Methods," in *Proceedings of the 4th Russian National Conference on Heat Exchange* (Moscow Power Engineering Institute Press, Moscow, 2006), Vol. 1, pp. 186–189 [in Russian].

21. O. A. Evtikhieva, B. S. Rinkevichyus, and A. V. Tolkachev, "Visualization of Nonstationary Convection in Liquids near Heated Bodies with Structured Laser Radiation," *Vestnik MEI*, No. 1, pp. 65–75 (2007).

22. I. V. Lebedev, B. S. Rinkevichyus, and E. V. Yastrebova, "Measuring local velocities of small-scale flows using a laser," *Journal of Applied Mechanics and Technical Physics*, **10**(5), pp. 805–807 (1969).

23. A. I. Imshenetsky, "Development and Calculation of Optico-Electronic Systems for Fluid Flow Diagnostics," Candidate's Dissertation (Moscow Power Engineering Institute, Moscow, 2005) [in Russian].

Chapter 7
Digital Refractogram Recording and Processing

7.1 Requirements for Image Recording and Processing Systems

As noted in Chap. 6, laser refractograms in the form of a two-dimensional spatial illumination distribution formed on a ground glass screen by the incident structured laser radiation passing through the medium under study are recorded with a digital photo or video camera.

The requirements for the characteristics of the camera and its software are primarily determined by the actual tasks carried out by the laser refractographic measuring system. The wide range of problems that are being solved by measuring the refractive index gradient of transparent media naturally prevent formulating requirements for a universal laser optoelectronic measuring system. But if one restricts oneself solely to the laboratory refractographic systems intended for studying optically inhomogeneous media, one can, to a certain degree of confidence, evaluate the requirements for their photoelectric part. In so doing, one should proceed from the general characteristics and parameters of the optical signals figuring in refractography. The signal being recorded here is the two-dimensional (in rectangular coordinates) distribution of illumination in the plane of the ground glass screen that is produced by laser radiation passing through the medium of interest.

The list of the main characteristics and parameters of the radiation recorded at the exit from the medium under study includes:

- Dynamic range of illumination
- Spectral characteristic of radiation
- Time characteristics (illumination correlation time, pulsed or continuous-wave radiation source)
- Spatial characteristics (spatial spectrum, spatial correlation window)

The variation range of screen illumination in laser refractography usually does not exceed 30–50 dB. The probe radiation sources used include helium–neon, argon, ruby, neodymium, or semiconductor lasers operated in continuous-wave or pulsed modes. The time constant of the nonstationary processes being studied, which in-

B. S. Rinkevichyus et al. (eds.), *Laser Refractography*,
DOI 10.1007/978-1-4419-7397-9_7, © Springer Science+Business Media, LLC 2010

fluences the rate of change of the illumination distribution, ranges from a few milliseconds to a few tens of seconds. The spatial frequency of typical refractograms comes to 1–100 mm^{-1}.

These parameters prevent the traditional analog optical image recording devices from being used in laser refractographic systems. The data on the two-dimensional spatial distribution of the screen illumination and its variations can only be obtained with up-to-date digital video equipment. The quality of conversion of the spatial illumination distribution into a digital data array governs the possibility of obtaining the maximum attainable estimates of the optical parameters of radiation passing through the optically inhomogeneous medium under study.

7.2   Digital Laser Refractogram Recording Systems

7.2.1   Main Characteristics of Refractogram Recording Systems

The outfit of a special-purpose digital laser refractogram recording system includes a photoelectric transducer, an analog-to-digital converter (ADC), and a computer–video interface. Software is an important element determining the quality of refractographic image recording and read-in.

At present, typical image recording systems make a most widespread use of photoelectric transducers in the form of matrix photo sensors built around charge-coupled devices (CCDs) [1, 2]. CCDs form the basis for constructing special-purpose photo and video cameras designed for carrying out scientific and technical tasks. Modern video cameras relying for their operation on CCDs find application in various measuring systems, optical spectrum analyzers, and contactless precision instruments for measuring coordinates, dimensions, distances, etc. The widespread use of CCDs is conditioned by such properties inherent in solid-state semiconductor devices as rigid raster, accurate tying of the geometrical coordinates of optical images and the time coordinates of their records in the video signal, possibility of preliminary signal processing in the CCD structure proper, linearity of the light-transfer characteristic, low supply voltages, high reliability, and also other valuable properties. An extremely important virtue of CCDs is their practically ideal PC compatibility, which makes it possible to program control the operation of the video camera, adapting it to various conditions, and record video signals on the hard disk of a computer via a standard interface.

The performance of modern professional video systems built around CCDs allows them to satisfy the above requirements for the recording of refractograms in the form of a two-dimensional screen illumination distribution. With the bit capacity of the analog-to-digital converter at the output of the CCD amounting to 12 bits, the dynamic range of the optical signal being recorded is over 60 dB. The resolution of standard video cameras ranges from 582×782 to $2,048 \times 2,048$ pixels, that of special-purpose ones reaching as high a value as 4,096×4,096 pixels. The exposure time can be varied from a few microseconds to a few tens of minutes (provided that external cooling is available). There is a possibility to program control the amplification of the camera and the reading of images into the computer memory in real time.

All processes in the camera (storage of charge packets, conversion of the preceding frame into a digital data array, data transfer) can occur simultaneously, and so the recording and observation of video signals and their reading onto the computer hard disk can be carried out continuously in TV and frame-by-frame operation modes. The spectral characteristics of standard CCDs allow one to use them in combination with sufficiently easily available lasers.

7.2.2 Digital Photo Camera-Based Refractogram Recording System

The use of a digital video camera is expedient when it is necessary to obtain refractograms of fast processes or a video film of an experiment [3]. If the objective is to get images of illumination distribution at individual instants of time, a digital photo camera, whose cost and size are much smaller than those of a video camera, would suffice. A wide range of digital photo cameras differing in characteristics are available at present [4]. Most important in the recording of refractograms are such photo camera characteristics as the effective number of pixels, sensitivity, focal length of the objective lens, exposure range, and bit capacity of the ADC. Table 7.1 lists as an example the technical data of the model *Konica Minolta Dimage Z20* digital photo camera that was used in the experiments described in Chap. 6. The camera–computer connection is via USB port. Images can be displayed online and recorded by means of special software.

Table 7.1 Technical data of the model *Konica Minolta Dimage Z20* photo camera

CCD matrix	1/2.5 in.
Effective number of pixels	5,000,000
Sensitivity (ISO)	Automatic; equivalent 50, 100, 200, 320 ISO
Aspect ratio	4:3
Maximum aperture	f/3.2 (wide-angle lens), f/3.4 (telephoto lens)
Focal length	6–48 mm
Minimal focusing distance (from the CCD matrix)	0.57 m (wide-angle lens position)
	1.57 m (telephoto lens position)
	0.08–1.07 m (Super Macro mode)
Shutter	Electronic, mechanical
Exposure range	1/2,000–4 s
LCD	3.8 cm (1.5 in.), TFT, color
ACD bit capacity	10
File formats	JPEG, Motion JPEG (mov, less audio)

7.2.3 SONY Video Camera-Based Refractogram Recording System

Consider the main characteristics of the refraction image recording system built around the model *SONY DCR-VX-2000E* semi-professional digital video camera that was also used in the experiments described in Chap. 6. The main characteristics of the video camera are presented in Table 7.2.

Experience in using this camera to record refractograms has shown the following shooting modes to be most popular:

- Manual focus mode
- Progressive scan mode
- Diaphragm-priority mode
- Exposure-priority mode (recording of nonstationary processes)

The camera–computer interfacing is by means of the model *IEEE 1394* interface. Worthy of note among the major specificities of this interface are the following:

- Use of a serial bus instead of a parallel interface
- Hot plugging (swapping) support
- Supply of external devices via the *IEEE 1394 interface* cable
- High data transfer rate
- Possibility of composing networks of various devices widely differing in configuration
- Simple configuration and ample capabilities
- Asynchronous and synchronous data transmission support

The type of the video camera–computer image transfer provided by the *IEEE 1394* interface makes it possible to maintain the efficiency of the experiment at a high level, provided that the data coming from the camera are processed online.

Table 7.2 Technical data of the model *SONY DCR-VX-2000E* digital video camera

Type	Class High-End digital video camera
Light-sensitive element	CCD 3 × 1/3 in., 450,000 pixels
Objective lens	12× optical zoom; 6–72 mm. Aperture ratio f/1.6–f/2.4
Focusing	Automatic, manual
Digital magnification	48×
Image stabilization	Super Steady Shot (optical)
Illumination range	2 lx Night shooting–Super NightShot 0.1 lx
Maximum shutter speed	1/10,000 s
Record format	mini DV, JPEG

7.2.4   The model Videoscan-285/B-USB Digital Video System

The model *Videoscan-285/B-USB* special-purpose digital video system is intended for reading black-and-white images into the computer memory and finds application in various domains of science and technology [5].

The system comprises a photoelectric converter, a photoreceiver control circuit, an ADC circuit, and a video camera–computer interface circuit. The replacement objective lens used in the system is an important component that determines the quality of the images being obtained. A simplified block diagram of the system is presented in Fig. 7.1.

The optical-to-electric signal conversion is effected by means of a photosensitive circuit built around Type *ICX285 AL* CCD. It contains a photosensitive matrix area wherein a charge relief composed of isolated charge packets is formed under the action of optical radiation. The magnitude of the charge in each packet is determined by the illumination of the matrix area wherein the given packet is formed, the size of this area, and the charge accumulation time (exposure time). The number of matrix elements (pixels) in Type *ICX285 AL* CCD amounts to $1,392 \times 1,040$, the pixel size is 6.45×6.45 μm, and the photosensitive area measures 8.98×6.7 mm.

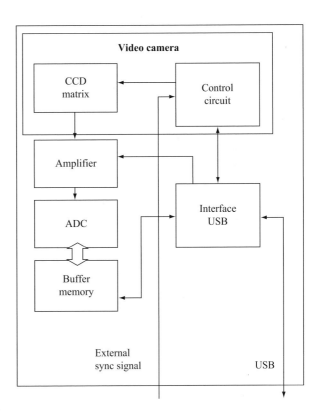

Fig. 7.1 Block diagram of the model *Videoscan-285/B-USB* video system

The photoreceiver control circuit produces pulse voltages that provide for the operation of the matrix photoreceiver elements; namely, charge accumulation and storage, charge transfer from the accumulation to the storage area, row-by-row charge transfer to the output device, and element-by-element reading of the charge packets. Once accumulated, all charge packets are transferred to the storage area wherefrom the elements of each row are read out consecutively.

The onset of the charge accumulation is determined by the type of synchronization used, which can be internal (a sync generator in the photoreceiver control circuit) or external (a computer program signal or an external sync signal). The output device that converts charge packets into a voltage across the load resistor allows integrating several pixels (2×2 or 4×4) directly in charge form. Such transformations augment in proportion the signal part of the charge packets and the maximum frame frequency of the camera, which naturally reduces its spatial resolution. Following the amplification of discrete analog readings and their conversion into digital form, the resulting digital data enter a buffer memory. When operating in the 8-bit mode, the relative output signal level of the camera ranges from 0 to 255 relative units.

The processes of charge accumulation, reading of the preceding frame, conversion of analog data into a digital array, and its entry into the computer via the USB port are conducted simultaneously. The operation of the system is controlled with special-purpose software.

The form of the spectral characteristic of Type *ICX285 AL* CCD is typical of silicon semiconductor substrates. It is presented in Fig. 7.2 in relative units (the characteristic is normalized to the maximum spectral sensitivity value). The spectral sensitivity of the video camera reaches its maximum at a radiation wavelength

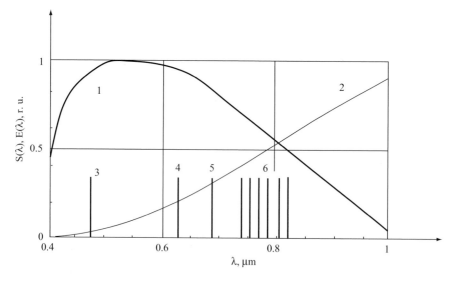

Fig. 7.2 Spectral characteristic $S(\lambda)$ of Type *ICX285 AL* CCD and radiation spectra $E(\lambda)$ of sources: *1*—spectral curve CCD; *2*—incandescent lamp; *3*—argon laser; *4*—He–Ne laser; *5*—ruby laser; *6*—semiconductor laser

Table 7.3 Comparative characteristics of the model *Videoscan-285/B-USB* system and the model *SONY DCR-VX-2000E* video camera

Characteristic	Videoscan-285/B-USB	SONY DCR-VX-2000E
Image forming device	1 black-and-white CCD matrix × 1,447,680 pixels	3 color CCD matrices × 450,000 (1,350,000) pixels
Image formats	1,392 × 1,040, 696 × 520, 348 × 260	880 × 228
Recording format	.bmp file	DV
Minimal illumination	–	2 lx (F1.6)
ADC bit capacity	8, 10	12
Charge accumulation time (exposure)	3.5 μs–10 min	0.1 ms–0.3 s
Sync modes	Internal, program, external	Internal, program, external
Scanning	Progressive	Progressive, interlaced
Frame frequency (Hz)	7.7	30
Shutter type	Electronic	Electronic
Resolution	500 TVL	530 TVL

around 0.5 μm and drops by a factor of 2 at wavelengths of 0.4 μm and 0.8 μm, which allows recording radiation of the argon, ruby, and helium–neon lasers.

Table 7.3 lists comparative characteristics of the model *Videoscan-285/B-USB* system and the above-considered the model *SONY DCR-VX-2000E* semi-professional digital video camera. The analysis of this table shows that as far as laser refractography tasks are concerned, the former system has a number of advantages over the latter camera. These include a substantially wider exposure range, a greater number of resolution elements, and the storage of images in the .bmp file format, which is necessary for the subsequent digital processing of refractograms.

While experimenting with this digital refractogram recording system, we determined a number of its characteristics, the lumination levels of the pixels of the image recorded by the camera being statistically averaged to calculate the mean value and variance of its output signal.

For the illumination distribution within the limits of the refractogram to be correctly represented, it is important that the illumination dependence of the output signal level of the camera be linear. This dependence was studied by varying according to a certain law the illumination level of a ground glass screen placed before the camera. The light modulator used was a polarizer that changed the transmission coefficient as a function of the tilt angle of the plate. Figure 7.3 shows the mean value of the recoded signal level as a function of the relative illumination level E_{rel} of the screen, plotted from experimental results. The analysis of this figure allows one to conclude that the light-transfer characteristic of the camera is sufficiently linear.

To properly select the image recording mode, we studied the effect of the exposure time T_{ex} and the amplification coefficient K_a of the electronic circuit on the parameters of the output signal of the camera. The exposure dependences of the mean value of the signal level recorded at various K_a values are presented in Fig. 7.4. At K_a values from 1 to 100, the plots are seen to be almost linear at exposures

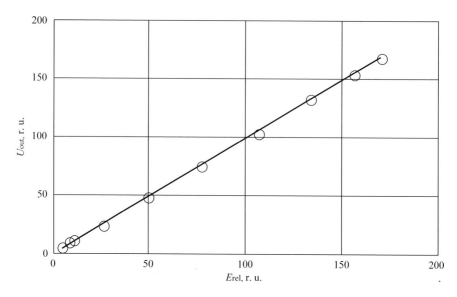

Fig. 7.3 Mean value of the recoded signal level as a function of the relative illumination level of the screen: *open circles*—experimental data points; *straight line*—approximation

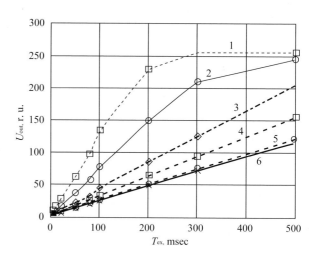

Fig. 7.4 Mean value of the output signal level as a function of T_{ex}: *1*—$K_a=300$; *2*—$K_a=200$; *3*—$K_a=100$; *4*—$K_a=50$; *5*—$K_a=10$; *6*—$K_a=1$

$T_{ex}<0.5$ s. At higher K_a values, saturation effects start coming into play, so that at $T_{ex}>0.2$–0.3 s the plots become substantially nonlinear. The results obtained are presented in Fig. 7.5 in the form of relationships between the mean output signal level and the amplification coefficient K_a at various exposure times T_{ex}. Figure 7.6 shows the effect of the exposure time T_{ex} on the root-mean-square (RMS) deviation of the output signal level. The analysis of these plots shows that as the exposure time grows longer, the RMS deviation of the output signal level increases, which is due to the effect of accumulation of noise of various origins. When recording weak

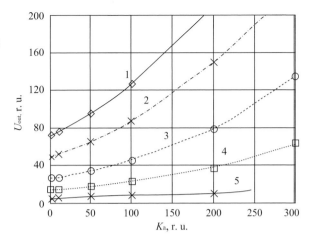

Fig. 7.5 Mean value of the output signal level as a function of K_a: 1—$T_{ex}=300$ msec; 2—$T_{ex}=200$ msec; 3—$T_{ex}=100$ msec; 4—$T_{ex}=50$ msec; 5—$T_{ex}=10$ msec

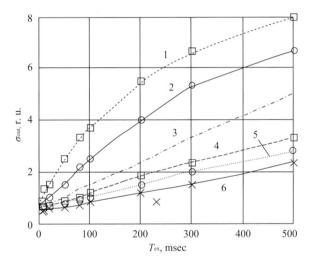

Fig. 7.6 RMS deviation of the output signal level as a function of the exposure time T_{ex}: 1—$K_a=300$; 2—$K_a=200$; 3—$K_a=100$; 4—$K_a=50$; 5—$K_a=10$; 6—$K_a=1$

optical signals, the presence of thermal generation noise proves important, its contribution to the image obtained growing higher with the increasing exposure time T_{ex}. Figure 7.7 shows the mean value and RMS deviation of the thermal generation noise level (dark operating mode of the camera) as a function of the exposure time T_{ex} at high T_{ex} values, when the relative level of the electric noise pickup from the operation of the electronic circuit drops and its effect on the resultant RMS deviation of the noise level weakens. These dependences were recorded with the camera operating in the 4×4 binning mode (joint signal accumulation by adjacent CCD cells) at an amplification coefficient of $K_a=100$. The analysis of the plots shows that the RMS deviation of the noise level is practically linear in T_{ex}, which agrees quite well with the notion of the origin of the thermal generation noise [2]. A typical histogram of the distribution of the relative noise levels U_n generated by the CCD

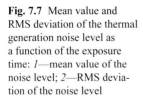

Fig. 7.7 Mean value and
RMS deviation of the thermal
generation noise level as
a function of the exposure
time: *1*—mean value of the
noise level; *2*—RMS deviation of the noise level

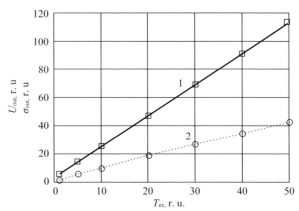

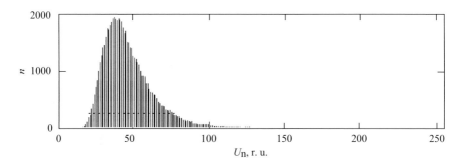

Fig. 7.8 Histogram of the distribution of the relative thermal generation noise levels (exposure
time $T_{ex}=20$ sec)

matrix cells $T_{ex}=20$ s is presented in Fig. 7.8. The thermal generation noise will
have a comparatively weak effect on the image being recorded if illumination is
strong and, accordingly, the exposure time, short. In such conditions, the major
contribution to the noise component of the output signal of the camera will be from
the shot noise. During the course of our experimental investigation, we obtained
numerical estimates of the variance of the output signal level of the camera at various illumination levels of the CCD matrix cells.

The study of the relationship between the variance of the output signal level
of the camera and the average illumination level of the CCD matrix cells showed
that under weak illumination conditions the variance of the noise component of the
output signal depended but little on the illumination level. As the illumination of
the CCD matrix was increased, the variance of the noise component of the output
signal grew higher practically linearly with increasing average output signal level.
These results confirm the conclusion that when the output signal is weak, the major contribution to its noise component comes from the signal-independent electric
noise pickup and thermal generation noise, and when the signal is strong, its noise
component is mainly contributed to by the shot noise of the optical radiation being
recorded.

Fig. 7.9 Two-dimensional distribution of the relative signal level over a frame fragment

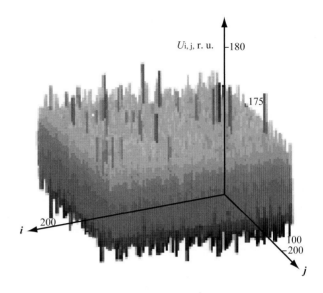

An important role in selecting an optimal refractogram processing algorithm is played by the information on the statistical characteristics of the CCD camera output signal under operating illumination conditions. Consider as an example a video image frame recorded under uniform background illumination conditions of the CCD matrix of the camera generating an output signal with an average relative level of $U_{out, av.} = 173$ relative units (r. u.). The plot of the two-dimensional distribution of the relative output signal level as a function of the row number i and the column number j of a frame fragment 256×256 pixels in size is presented in Fig. 7.9. The histogram of the distribution of the relative brightness values of the pixels within the boundaries of the whole frame is shown in Fig. 7.10. One can see that under sufficiently uniform illumination conditions, there is also a noise component within the boundaries of the frame. It follows from Fig. 7.11 that the distribution law of this component is close to the Gaussian law. Our analysis of the autocorrelation function

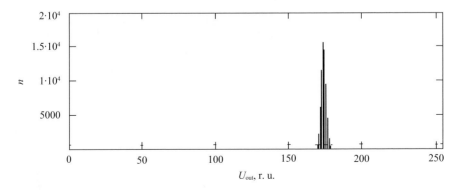

Fig. 7.10 Histogram of the relative signal level distribution within the boundaries of an image frame

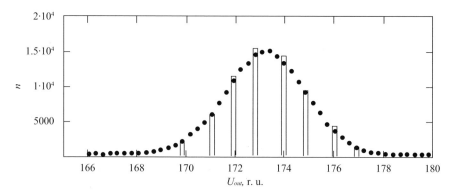

Fig. 7.11 Histogram of the relative signal level distribution in comparison with the normal distribution histogram: ПППП —signal level distribution histogram; •••—normal distribution histogram

of the frame fragment has shown that the signal levels in spatially separate image elements under uniform illumination conditions are practically uncorrelated.

Consider as an example a video image frame recorded in conditions of weak background illumination of the CCD matrix of the camera, where its average relative output signal level is $U_{out,\ av.} = 8.6$ r. u. The plot of the two-dimensional distribution of the relative signal level over a frame fragment 256×256 pixels in size is presented in Fig. 7.12. It can be seen that given this illumination level and the 8-bit operating mode of the ADC of the camera, the noise component is perceptibly contributed to by the quantization noise.

The characteristics of the image recorded are affected not only by noise of various origins (shot noise of the image recording process, thermal generation noise of

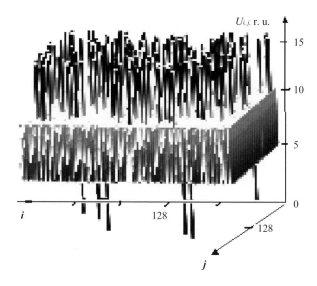

Fig. 7.12 Two-dimensional distribution of the relative signal level over a frame fragment under weak illumination conditions

Fig. 7.13 Two-dimensional spatial spectrum modulus of a frame fragment

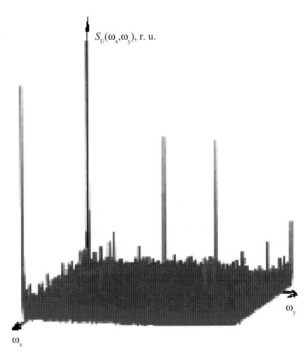

$S_U(\omega_x, \omega_y)$, r. u.

ω_y

ω_x

the CCD, quantization noise of the ADC), but also by the electric noise pickup from the electronic photoreceiver control circuit. Inasmuch as the main noise pickup sources are the pulse periodic line and frame scanning signals, the levels of these noise pickups can most readily be estimated in the spectral domain. The plot of the two-dimensional spatial spectrum modulus of a frame fragment recorded under operating illumination conditions is shown in Fig. 7.13. Standing out against the background of the uniform level of the noise spectral components in this plot are individual high-amplitude harmonics whose frequencies are multiples of the line and frame frequencies.

Experiments also showed that as the illumination level was reduced, the relative contribution from individual harmonic components of the spatial spectrum to the output signal increased perceptibly. The harmonic composition of the spatial spectrum of the noise pickup from the electronic circuit can be revealed most explicitly by analyzing an image frame obtained under dark operating conditions where there is no shot noise of the signal. The analysis of the spatial spectrum of the output signal under these operating conditions showed it to contain regularly arranged spatial spectrum harmonics with frequencies related to the line and frame frequencies of the camera.

Among important characteristics of an optoelectronic recording system are the resolving power (resolution) it offers and the estimated error of the image element displacement in the images obtained. For the measure of the resolving power of such a system, one can use either directly the length ΔL of the minimal line segment

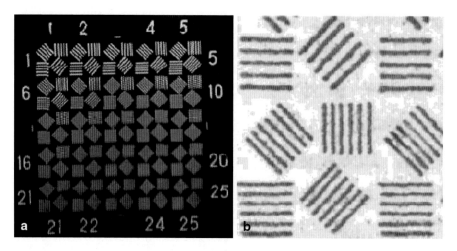

Fig. 7.14 Resolution test chart. **a** Full image of the chart. **b** Magnified images of the chart elements

between the lines being resolved in the image or its reciprocal determining the number of resolvable lines per millimeter. The resolution characteristics of the model *Videoscan-285/B-USB* video recording system using the model *AVENIR SE5018* objective lens with a focal length of 50 mm and a relative aperture from 1:1.8 to 1:16 were evaluated experimentally by means of resolution test charts (patterns) [6]. A resolution test chart is a transparent plate with a table comprising 25 elements (large squares) differing in line pitch applied to it (Fig. 7.14).

The main characteristics of resolution test charts are listed in Table 7.4. Each test chart is characterized by its base *B*—the distance specified with high precision between two marks made at the sides of the chart (see Fig. 7.14). The number of lines per millimeter for the *i*th element of the chart, R_i, is specified in terms of the chart base *B* by the formula

$$R_i = \frac{60}{B} k_i, \tag{7.1}$$

where $k_i = (1.06)^{i-1}$.

The values of the quantity k_i are listed in Table 7.5.

Table 7.4 Characteristics of resolution test charts

Test chart No.	Chart base (mm)	Number of lines per 1 mm	
		Maximum (element No. 25)	Minimum (element No. 1)
1	1.2	200	50
2	2.4	100	25
3	4.8	50	12.5
4	9.6	25	6.5
5	19.2	12.5	3.1

Table 7.5 Values of the quantity k_i

i	1	2	3	4	5	6	7	8	9	10	11	12	13
k_i	1	1.06	1.12	1.19	1.26	1.34	1.42	1.51	1.6	1.7	1.8	1.9	2.01
i	14	15	16	17	18	19	20	21	22	23	24	25	
k_i	2.13	2.26	2.4	2.54	2.59	2.85	3.02	3.20	3.4	3.6	3.8	4.0	

The estimation by the Rayleigh criterion of the resolution of the optical system under study at an objective lens-to-test chart distance of $s=25$ cm showed that the maximum number of resolvable lines per millimeter (in the object plane) corresponded to element No. 25 of test chart No. 5; i.e., $R_{25}=12.5$. The minimum resolvable line pitch in that case was 80 μm. This estimate allows one to conclude that the minimum displacement of refractogram elements that can be measured in this system also amounts to 80 μm.

The line pitch in the images of test chart elements was estimated by a spectral method. The theoretical line pitch values in the test chart elements, the experimentally measured average line pitch values in the images of the test chart elements, and the resulting magnification constant values of the optical system are listed in Table 7.6.

The estimated magnification constant Γ of the optical system, found from measurements taken at an objective lens-to-test chart distance of $s=25$ mm and an objective lens focal length of $f=50$ mm, amounts to some 0.25. This is an approximate value, for there are no data available on the exact coordinate of the front principal plane of the objective lens. Nevertheless, the magnification constant estimates are sufficiently close to one another (the relative RMS deviation is a mere 1.8%) and approximately correspond to the theoretical value.

To estimate errors in measuring spatial displacements of refractogram elements, experiments were conducted on measuring the displacements a resolution test chart was made to move through by means of a micrometric displacement device. The processing algorithm consisted in estimating the displacement of the image of a small square of the chart from the phase difference between the first harmonics of the Fourier spectra of the one-dimensional spatial intensity distributions of the original and displaced images averaged over the x-axis. The computations were made using the magnification constant values listed in Table 7.6. The experimental results are presented in Table 7.7.

The analysis of these results shows that the image displacement values estimated by processing the data for the individual image elements differ by no more than 1%, which bears witness to the fact that the random measurement error component is

Table 7.6 Line pitch and magnification constant values

Test chart element No.	11	16	21	25
Theoretical line pitch (μm)	178	133	100	80
Measured average line pitch in the image of the test chart element (μm)	45.5	34.1	24.5	19.9
Estimated magnification constant β	0.256	0.255	0.245	0.249

Table 7.7 Results of resolution test chart displacement measurements

Induced test chart displacement (μm)	50				500			
Test chart element No.	11	16	21	25	11	16	21	25
Estimated image displacement (μm)	11.8	12.2	11.9	12.2	122	122	121	123
Estimated test chart element displacement (μm)	46.3	47.7	48.5	48.9	477	478	495	495

small. The estimated systematic error component in measuring test chart displacements for chart elements Nos. 21 and 25 is 5 μm, which is also around 1%. Note that this error value proves materially smaller than the estimated resolution of $R = 80$ μm found earlier to limit the accuracy of measuring the displacements of refractogram elements. The results presented do not contradict one another, for the test chart displacements were measured by processing not a single line in the test chart image but several spatially periodic lines contained in a single element (small square) thereof.

The experimental investigation of the optoelectronic digital refractogram recording system built around the model *Videoscan-285/B-USB* matrix CCD video camera equipped with the model *AVENIR SE5018* objective lens allows one to draw a number of conclusions:

1. Recording optical radiation under controllable illumination conditions makes it possible to estimate the linearity region of the light-transfer characteristic of the camera and properly select its magnification constant and exposure time.
2. The exposure dependences of the mean value and RMS deviation of the noise level at the output of the CCD camera under dark operating conditions are of linear character for long exposures, which confirms the conclusion that the thermal generation noise in these conditions is predominant, the relative level of the electric noise pickup from the operation of the electronic circuit and its effect on the resulting RMS deviation of the noise level being reduced.
3. If the illumination of the object being observed is high and, correspondingly, the exposure time short, the major contribution to the noise component of the output signal comes from the shot noise of the recorded optical radiation, the variance of the output signal of the camera being proportional to the illumination of the object.
4. Under area uniform and stationary illumination conditions of the object of observation, the distribution law of the camera output signal level (averaged over the pixels of the image, with level quantization disregarded) is close to the Gaussian law, and the two-dimensional spatial autocorrelation function (ACF) of the signal has a pronounced peak in the region of small displacements.
5. When the illumination of the object is weak and the exposure time is not very long, the relative contribution to the output signal of the camera from individual harmonic components of the spatial spectrum, due to the effect of the noise pickup from the electronic circuit, proves substantial and manifests itself as a periodic space modulation of the ACF signal level. The space–time and spec-

tral characteristics of the noise pickup from the electronic circuit depend on the operating conditions of the camera (one-time or continuous recording, frame frequency, exposure time, binning present or not, etc.), type of the camera–computer interface hardware, degree of screening and length of the connecting cable, specific features of the electronic camera control circuit, and design of the electronic block. To effectively filter out the noise induced by the electronic circuit, the noise characteristics should preliminarily be analyzed for each particular camera under operating conditions.

6. When the illumination is weak and the ADC of the camera is operated in the 8-bit mode, quantization noise contributes perceptibly to the noise component of the output signal.

7. The method worked out for estimating the resolution of the optoelectronic recording system and the error in measuring the displacements of image elements by means of test charts made it possible to establish that for the given recording system the minimal distance between resolvable elements of the test patterns and displacement estimation error proved smaller than the expected spatial variations of refractograms, which allows one to conclude that this system can be used in laser refractography practice.

8. The estimates presented above for the distribution law parameters of the energy spectra and correlation functions of the processes at the output of the recording system, as well as those for the resolution of the system and the test chart displacement measuring error, make it possible to approximately determine the expected errors in measuring the characteristics of the refractograms being recorded and suggest processing algorithms improving the accuracy of such estimates.

9. The method developed for investigating the characteristics of CCD video cameras can be used to analyze other digital image recording systems.

7.3 Digital Laser Refractogram Models

A laser refractogram is a two-dimensional illumination distribution produced on a ground glass screen by structured laser radiation passing through the medium under study. A most simple model of an illumination distribution undistorted by the medium, which is most adequate to the majority of practical applications, is the cross section of an astigmatic laser beam ("laser plane") by the plane of the screen. This is an area bounded by an ellipse with half-axes differing substantially in size. The image of the illuminated screen is recorded with the above-described hardware.

The processing of refractograms consists in determining the difference between the spatial coordinates of various elements of the two-dimensional illumination distribution in the plane of the screen, measured in the absence of the medium under study and in its presence. The spatial changes in the illumination distribution carry information about the physical parameters of the medium.

In the preceding sections (Chaps. 4–6), we have presented examples of refractograms obtained through computations by mathematical models and experimentally. The computations made in the geometrical optics approximation have allowed us to obtain the pictures of distortions of the mathematical model of an infinitely "thin" laser plane whose undistorted trace on the screen is a line segment. Analyzing experimental results makes it possible to draw conclusions as to the specific features of the formation process of actual refractograms, not allowed for in the theoretical models.

A physically realizable laser plane is always of finite width; the power density of laser radiation is distributed in space by a certain distribution law and is prone to random fluctuations. The recording of a laser plane with a digital video camera gives rise to additive noise. These factors should be taken into consideration in selecting the mathematical models of typical refractograms that are necessary for comparing between theoretical and experimental results and also for working out and testing the processing algorithms considered in the next chapter.

The numerical refractogram models were implemented in the MATLAB program package. When creating the models, it was assumed that the recording device was a matrix-type CCD photoreceiver whose main characteristics have been considered earlier in the text. The model parameters were selected to be as follows:

- Number of pixels in the CCD matrix along the horizontal, M, and along the vertical, N
- Dark current parameters of the photo camera
- Bit capacity (8, 10, or 12) of the ADC of the photo camera
- Physical size of the CCD matrix cell along the x-axis, l_x, and along the y-axis, l_y

It was assumed in the model that the gaps between the cells in the CCD matrix were negligibly small and the cells themselves had the same geometrical dimensions in the rows and columns. It was assumed that the CCD matrix under consideration had a linear light-transfer characteristic and that the useful output signal was proportional to the illumination of the respective pixel of the CCD matrix. The total illumination of each pixel was determined as the sum of the useful component, the noise component, and the background component. The dark current of the CCD matrix, caused by the thermal generation of charge carriers during exposure, was taken to be described by a normal distribution with specified parameters, such as mathematical expectation and variance. The matrix of the noise component n was formed with a random number generator with a normal distribution and specified mathematical expectation and variance. For simplicity, the background in the model was specified assuming a constant illumination level of the entire matrix. Thus, the relative signal level of each pixel, corresponding to the brightness $U_{i,j}$ of the elements of the image being recorded, was defined as follows:

$$U_{i,j} = g\{H_{i,j} + n_{i,j} + C\}, \tag{7.2}$$

$$H_{i,j} \sim F(x_i, y_j),$$

where i is the serial number of the row of the image, j is that of the column of the image, x and y are the spatial rectangular coordinates, $F(x, y_j)$ is the illumination of the CCD matrix cell in the position i, j, $H_{i,j}$ is the useful signal, $n_{i,j}$ is the noise component, C is the background level, and g is the level quantization operator.

When refractograms are recorded with a digital photo camera, a spatial quantization of the image and a recording of a two-dimensional array of samples obtained by integrating illumination over the area of the respective photosensitive cell of the CCD matrix over the time of exposure take place. Thus, a refractographic image is represented by an M by N numerical matrix wherein the value of each element corresponds to a certain quantization level of its energy characteristic—illumination (pixel coordinate system) [7].

In the numerical model, the illumination of each pixel of the image was represented by an integral 16-bit number. The image could be stored on a hard disc in the .tif format allowing for a depth of change in the pixel illumination of up to 16 bits.

Let the continuous illumination distribution in the plane of recording be $F(x, y)$ and let the center of the photoreceiver matrix coincide with the optical axis of the system. In that case, the level of the video signal specifying the brightness of the image pixel with the serial number (i, j), $i=0,\ldots, N-1, j=0,\ldots, M-1$, is defined by the relation

$$U_{i,j} = k \cdot T_{\text{ex}} \int_{\xi_i - 0.5l_x}^{\xi_i + 0.5l_x} \int_{\eta_j - 0.5l_y}^{\eta_j + 0.5l_y} F(\xi, \eta) d\xi d\eta,$$

where k is the normalizing factor, T_{ex} is the exposure time,

$$\xi_i = \xi_{\max} - (i + 0.5)l_x, \quad \xi_{\max} = \frac{N}{2}l_x,$$

$$\eta_j = \eta_{\min} + (j + 0,5)l_y, \quad \eta_{\min} = -\frac{M}{2}l_y.$$

The relations defining the coordinates of the CCD matrix cells are illustrated by Fig. 7.15.

Most simple to process is the image of the screen illumination distribution for a probe radiation in the form of a laser plane in the absence of the medium under study. In this case, the two-dimensional illumination distribution in the CCD matrix plane may be described by the model of the cross-section of an astigmatic Gaussian beam with dimensions $w_x \gg w_y$:

$$F(x, y) = F_0 \exp\{-(x_{0n} - x_n)^2 / w_x^2\} \cdot \exp\{-(y_{0n} - y_n)^2 / w_y^2\}, \quad (7.3)$$

$$x_n = x \cdot \cos\alpha + y \cdot \sin\alpha, y_n = -x \cdot \sin\alpha + y \cdot \cos\alpha,$$

$$x_{0n} = x_0 \cdot \cos\alpha + y_0 \sin\alpha, y_{0n} = -x_0 \cdot \sin\alpha + y_0 \cdot \cos\alpha,$$

Fig. 7.15 Coordinates of the
CCD matrix cells

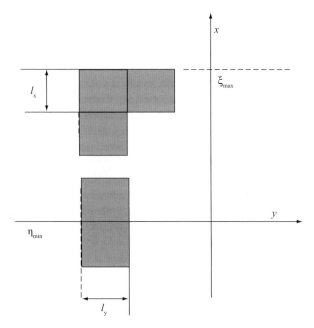

where F_0 is the illumination at the center of the beam cross-section; x_0 and y_0 are the coordinates of the beam cross-section center in the coordinate system x, y; x_{0n} and y_{0n} are the coordinates of the beam cross-section center in the coordinate system x_n, y_n; w_x and w_y are the dimensions of the beam cross-section; α is the angle of incli-nation of the laser plane image to the x-axis, and x, y, x_n, y_n are coordinates in the coordinate systems arranged with respect to the CCD matrix as shown in Fig. 7.16. Conversion to the new coordinates x_n, y_n is required for modeling the rotation of the laser plane through the angle α.

The power density F_0 at the center of the beam cross-section is related to the laser power P by the relation [8]

$$F_0 = \frac{2P}{\pi \times w_x \times w_y}. \qquad (7.4)$$

Thus, one can select the following parameters for the useful signal model:

x_0, y_0 —coordinates of the beam cross-section center;
w_x and w_y —dimensions of the beam cross-section along the x- and the y-axis, re-spectively;
α—angle of inclination of the laser plane to the x-axis;
A—value of the useful signal level at the center of the Gaussian beam, proportional; to the illumination F_0, exposure time, and the area of the photosensitive cell.

Relations (7.2)–(7.4) define the parametric model of the refractogram of the probe radiation in the form of a laser plane in the absence of the medium being studied.

Fig. 7.16 Mutual arrangement of coordinate systems and the CCD matrix

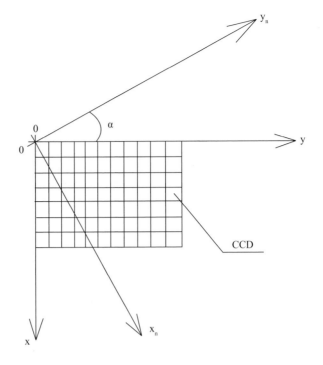

The model takes account of the laser beam parameters and the characteristics of the digital refractogram recoding system. The effect of the medium under study will manifest itself in changes in one or several parameters of the given model, which are to be determined in the course of processing of model or experimental refractograms. Some exemplary models of images formed when illuminating a screen with a laser plane are presented in Fig. 7.17. Figure 7.18 shows a two-dimensional spatial illumination distribution in the image of a refractogram.

In a number of cases, it proves possible to construct a parametric refractogram model for a medium showing a radially symmetric refractive index distribution on being probed with a laser plane. This can be exemplified by the refraction patterns obtained experimentally with a laser plane passing beneath a hot ball cooling in water (see Chap. 5). To describe such a model, it proves convenient to introduce the

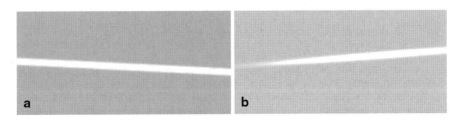

Fig. 7.17 Model laser plane images differing in inclination angle. **a** Negative inclination angle. **b** Positive inclination angle

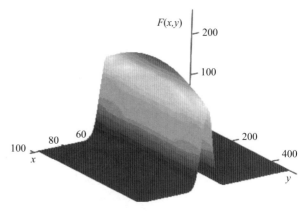

Fig. 7.18 Isometric drawing of a model laser plane image

notion of the central line of the screen illumination distribution, meaning thereby an imaginary curve formed by the intersection of an infinitely thin laser plane passing through the medium under study and the plane of the screen. Subject to certain conditions, such a line in the refraction region can be approximated in the rectangular coordinate system of its own by an ellipse with the following parameters: a and b—the dimensions of the ellipse half-axes, x_0 and y_0—the coordinates of the ellipse center, and φ—the angle of inclination of the major half-axis of the ellipse to the x-axis. The position of the coordinate system in the plane of the screen is determined by the distance y from the origin to the straight line passing through the unperturbed sections of the trace.

However, parametric screen illumination distribution models that carry information about the extent and character of the inhomogeneities of the optically transparent medium under study are not always possible to construct. Refractograms frequently have a complex shape that is difficult to describe analytically. In such cases, theoretical laser refractograms can be represented by their numerical models, and their comparison with experimental results should be made by one of the generally accepted methods for estimating approximation errors (least squares method, least modules method, etc.) [9].

7.4 Digital Refractogram Processing Methods

7.4.1 Parametric Estimation Methods

Parametric estimation methods provide for the approximation of the experimental function being processed by an analytical model containing a set of informational and accompanying parameters. The model parameters are estimated by one of the criteria selected.

The active window of the program allowing for the approximation of experimental refractographic images by an elliptical model is shown in Fig. 7.19. Approxima-

Fig. 7.19 Active window of
the program for determining
elliptical laser refractogram
model parameters

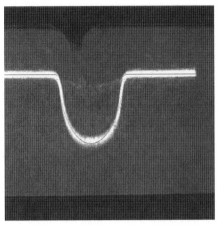

Parameters
of elliptic model

a: 86

b: 221

x0: 417

y0: 108

Fi: 0

y: .55

tion can be carried out by manual or automatic selection of model parameters that
most faithfully describe the central line of the experimental illumination distribution being processed. The automatic determination of the model parameters is effected by the method of exhaustive search for the maximum sum of illuminations
of the pixels of the smoothed experimental image, whose coordinates coincide with
those of the respective points in the elliptic model.

Additional functions of the program include the construction of plots for the illumination distribution in the desired column of the CCD matrix (along the x-axis),
or for the distribution averaged over several columns, and also the estimation of the
parameters of the Gaussian model of the distribution selected. Subject to processing here are red-, green-, and blue-colored experimental refractograms, as well as
black-and-white images differing in the number of gray gradations. The appearance
of the window of the additional functions of the program is shown in Fig. 7.20.
The color to be processed is selected in the window located on the left below the
distribution histogram. To the right of this window is an entry field that can help
to approximately estimate the signal and noise levels in the desired column of the
image. Arranged in the right-hand part of the window of additional functions are

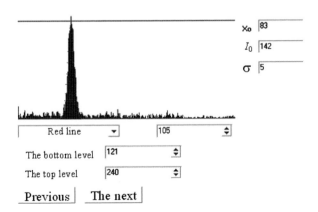

X_0 83

I_0 142

σ 5

Red line   105

The bottom level   121

Fig. 7.20 Appearance of
the illumination distribution
estimation window

The top level   240

Previous   The next

control buttons and a display window for the results of estimation of the Gaussian illumination distribution model parameters X_0, I_0, and σ, where X_0 is the serial number of the distribution center row, I_0 is the amplitude, and σ is the distribution width at the I_0/e level. In the example presented in Fig. 7.20, the center of the Gaussian distribution is located in the row $X_0 = 83$ and has a maximum relative illumination level of $I_0 = 142$ and an effective distribution width of $\sigma = 5$.

A similar Gaussian distribution can also be used to approximate the screen illumination distribution image in the absence of the medium distorting the shape of the laser plane. By processing this image, one can determine the model parameters, such as the angle α of the laser plane inclination to the x-axis and the coordinates x_0 and y_0 of the laser planer center.

To estimate these parameters, use can also be made of the algorithm based on the weighting method [9]. The essence of this method consists in "weighting" the discrete signal in each column of the image within a window of specified size in the neighborhood of the point where the illumination of the screen by the astigmatic laser beam is at its maximum. Qualified coordinates of the local maxima are used in estimating the parameters of the straight line approximating the laser plane, namely, the inclination angle α and the coordinates x_0 and y_0 of the center of the Gaussian beam, by the minimum RMS deviation method [10].

The coordinates of the laser plane center in each column of the image are defined by the relations

$$\xi_j = \frac{\sum\limits_{i=q-\frac{r}{2}}^{q+\frac{r}{2}} \sum\limits_{v=j}^{j+s} iv U_{i,v}}{\sum\limits_{i=q-\frac{r}{2}}^{q+\frac{r}{2}} \sum\limits_{v=j}^{j+s} v U_{i,v}}; \quad \zeta_j = \frac{\sum\limits_{i=q-\frac{r}{2}}^{q+\frac{r}{2}} \sum\limits_{v=j}^{j+s} iv U_{i,v}}{\sum\limits_{v=j}^{j+s} \sum\limits_{i=q-\frac{r}{2}}^{q+\frac{r}{2}} i U_{i,v}}, \tag{7.5}$$

where $U_{i,j}$ is the illumination level of the pixel numbered i,j, q is the serial number of the row accommodating the image element $U_{q,j}$ having the maximum illumination within the jth column, and $r \times s$ are the dimensions of the weighting window. The points with the coordinates ξ_j and ζ_j, where $j = 1, \ldots, (M-s)$, are used to estimate the parameters p_1 and p_2 of the straight line that can be used to approximate the laser plane:

$$\xi = p_1 \times \zeta + p_2,$$

$$p_1 = \frac{(M-s) \times \sum\limits_{j=1}^{M-s} (\xi_j \cdot \zeta_j) - \sum\limits_{j=1}^{M-s} \xi_j \times \sum\limits_{j=1}^{M-s} \zeta_j}{(M-s) \times \sum\limits_{j=1}^{M-s} \zeta_j^2 - \left(\sum\limits_{j=1}^{M-s} \zeta_j\right)^2},$$

$$p_2 = \frac{1}{(M-s)} \times \left(\sum\limits_{j=1}^{M-s} \xi_j - p_1 \times \sum\limits_{j=1}^{M-s} \zeta_j\right). \tag{7.6}$$

To estimate the parameters p_1 and p_2, use is made of the minimum RMS deviation method [10]. The estimates obtained for the parameters of the approximating line are used to determine the laser plane parameter estimates $\widehat{\alpha}$ and $\widehat{x}_0$:

$$\widehat{\alpha} = \arctan(p_1), \quad \widehat{x}_0 = p_2 - \tan(\widehat{\alpha})y_0, \tag{7.7}$$

where y_0 is the second coordinate of the center of the Gaussian beam.

The algorithm considered above makes it possible to estimate the parameters of laser planes inclined to the x-axis at angles α in the range from $-45°$ to $+45°$. Supplementing this algorithm with the function of determining the image scanning direction allows processing images of a screen illuminated with a laser plane inclined at an angle over $45°$.

The above-described model of a refractogram appearing on a ground glass screen illuminated with a laser plane was used in the program written in the MATLAB technical computing language [11].

On starting the program, one selects the model parameters such as the mathematical expectation and variance of noise, image size in pixels, laser beam dimensions w_x and w_y, coordinates x_0 and y_0 (in pixels) of the beam center, beam inclination angle α, illumination A (in relative units of gray level values) produced by the beam at the center, and the number of images with the given parameters to be modeled. The program is capable of processing both model and experimental images downloaded from an external file. An example of the display of a dependence selected by the user is presented in Fig. 7.21.

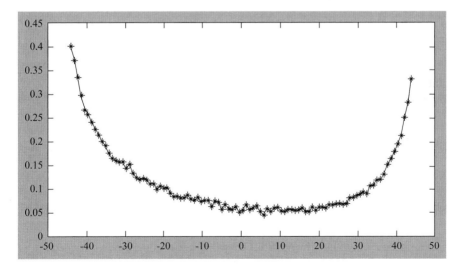

Fig. 7.21 Display window for the results of the error analysis of refractogram parameter estimates

7.4.2 Weighting Algorithm Error Analysis

Let us present some results of the analysis of errors in determining the parameters of a refractogram obtained when illuminating a screen with a laser plane. The RMS deviations of the estimates of the center coordinates of the refractogram and its inclination angle are defined by the relations [12]

$$\sigma[\hat{x}_0] = \sqrt{\overline{\hat{x}_0^2} - \left(\overline{\hat{x}_0}\right)^2}, \ \sigma[\hat{\alpha}] = \sqrt{\overline{\hat{\alpha}^2} - \left(\overline{\hat{\alpha}}\right)^2}, \tag{7.8}$$

where $\overline{()}$ stands for the mathematical expectation computing operation. The mathematical expectation estimate was taken to be the sample mean μ from $m=30$ model experiments:

$$\mu(x_0) = \frac{1}{m} \cdot \sum_{i=1}^{m} (\hat{x}_0)_i, \ \mu(\alpha) = \frac{1}{m} \cdot \sum_{i=1}^{m} (\hat{\alpha})_i. \tag{7.9}$$

The systematic errors of the estimates were found from the relations [12]

$$b[\hat{x}_0] = \mu(\hat{x}_0) - x_0, \ b[\hat{\alpha}] = \mu(\hat{\alpha}) - \alpha. \tag{7.10}$$

By varying the model parameters, one can analyze the errors of its parameter estimates.

Table 7.8 lists the systematic errors in estimating the center coordinates and inclination angle of the refractogram for the following model parameters: $A=240$ r. u. (gray level values), $\alpha=30°$, $w_y=2,000$ pixels, $w_x=4$ pixels, an RMS deviation of the noise level of 4 r. u., a background brightness of $C=0$, and a window size of 8 pixels.

It can be seen from the data of Table 7.8 that to record refractograms, it is expedient to use a CCD with a large number of pixels.

To automatically process refractograms of intricate shape (such as containing loops or bends), the full frame is often required to be broken down into regions of a smaller size. It is, therefore, of practical interest to analyze the effect of the size of such regions on the errors of the refractogram parameter estimates.

Table 7.8 Systematic errors in estimating the center coordinates and inclination angle of the refractogram for a laser plane

Number of image elements	32 × 32	64 × 64	128 × 128	256 × 256	512 × 512	1024 × 1024
Center coordinates (pixel)	$x_0=16;$ $y_0=16$	$x_0=32;$ $y_0=32$	$x_0=64;$ $y_0=64$	$x_0=128;$ $y_0=128$	$x_0=256;$ $y_0=256$	$x_0=512;$ $y_0=512$
$b[x_0]$ (pixel)	0.247	0.218	0.013	<0.001	<0.001	<0.001
$b[\alpha]$ (deg.)	0.243	0.022	0.008	<0.001	<0.001	<0.001

The analysis of the relationship between the bias of the estimate of the coordinate x_0 and the inclination angle of the laser plane for the model parameters $M=1{,}024$ pixels, $N=1{,}024$ pixels, $w_y=2{,}000$ pixels, $w_x=2$ pixels, $A=240$ r. u., $C=0$ (no background), and a window size of 4 pixels in the absence of noise (methodic error) has shown that as the inclination angle of the laser plane increases, so does the bias of the estimate. At the same time, the absolute value of the bias does not exceed a few hundredth fractions of a pixel, and so one can conclude that the methodic error of the coordinate estimate can be disregarded even in cases where the size of the observation window is small. The bias of the estimate of the inclination angle α varying over a wide range also proves negligible (on the order of 10^{-3} deg.).

The methodic components of the refractogram parameter estimates, obtained with the use of the given models, are negligibly small, because these models fail to allow for the noise present in actual refractograms. Figures 7.22 and 7.23 present the respective characteristics of the estimates $\widehat{x}_0$ and $\widehat{\alpha}$ as a function of the inclination angle of the refractogram for a model taking account of the effect of noise, wherein $M=32$ pixels, $N=32$ pixels, $w_y=2{,}000$ pixels, $w_x=4$ pixels, $A=240$ r. u., the noise RMS deviation is 4 r. u., $C=0$ (no background), and the size of the window is 8 pixels. One can see from Fig. 7.22 that the bias of the center coordinate estimate in this case reaches a few tenth fractions of a pixel and grows higher with increasing inclination angle. The bias of the estimate of the angle α behaves similarly, reaching a few degrees. The RMS deviations of the estimates prove substantially smaller.

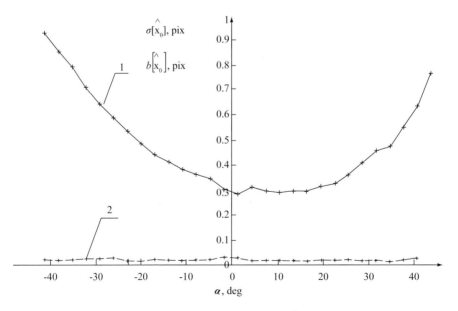

Fig. 7.22 Statistical characteristics of the coordinate estimate $\widehat{x}_0$ as a function of the inclination angle α: $M\times N=32\times32$; 1—bias of the estimate; 2—RMS deviation of the estimate

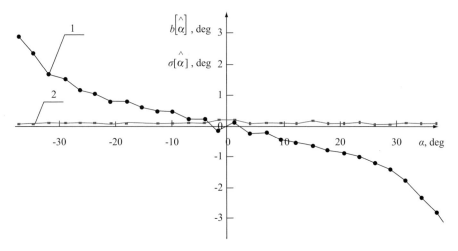

Fig. 7.23 Statistical characteristics of the inclination angle estimate $\widehat{\alpha}$ as a function of the magnitude of this angle: $M{\times}N{=}32{\times}32$; *1*—bias of the estimate; *2*—RMS deviation of the estimate

Presented in Figs. 7.24 and 7.25 are the inclination angle dependences of the errors of the estimates $\widehat{x}_0$ and $\widehat{\alpha}$ at enlarged frame dimensions: $M{=}256$ pixels, $N{=}256$ pixels, $w_y{=}2{,}000$ pixels, $w_x{=}4$ pixels, $A{=}240$ r. u., the RMS deviation of the noise level is 4 r. u., $C{=}0$ (no background), and the size of the window is

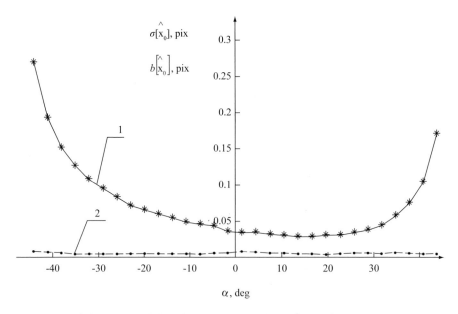

Fig. 7.24 Statistical characteristics of the coordinate estimate $\widehat{x}_0$ as a function of the inclination angle α: $M{\times}N{=}256{\times}256$; *1*—bias of the estimate; *2*—RMS deviation of the estimate

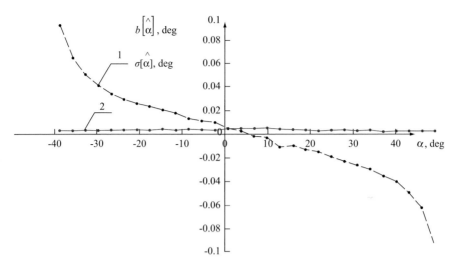

Fig. 7.25 Statistical characteristics of the inclination angle estimate $\hat{\alpha}$ as a function of the magnitude of this angle: $M \times N = 256 \times 256$; *1*—bias of the estimate; *2*—RMS deviation of the estimate

8 pixels. Comparing the plots in these figures with their counterparts presented in Figs. 7.22 and 7.23 shows that the errors decrease with increasing frame dimensions. At the same time, owing to the edge effects, the bias of the estimates increases at inclination angles over $40°$ and under $-40°$.

When the laser power P in the model is fixed, so that the image parameter A becomes a dependent variable defined by relation (7.4), the relationships between the estimates of the parameters being computed and the dimensions w_y and w_x turn out to be practically important. The results of the analysis of the statistical characteristics of the error of the coordinate estimate $\hat{x}_0$ as a function of the dimension w_x of the Gaussian laser beam at $M = 256$ pixels, $N = 256$ pixels, $x_0 = 128$ pixels, $y_0 = 128$ pixels, $w_y = 2{,}000$ pixels, $\alpha = 0°$, an RMS deviation of the noise level of 4 r. u., $C = 0$ (no background), and a window size of w_x are presented in Fig. 7.26. It can be seen that as the dimension w_x is enlarged, the RMS error of the estimate of the parameter x_0 increases but insignificantly. This is due to the fact that enlarging w_x with the laser power remaining invariable reduces the illumination of the CCD matrix, which in turn reduces the signal-to-noise ratio in the refractogram.

Figure 7.27 presents the statistical characteristics of the error of the estimate of the parameter x_0 as a function of the dimension w_y of the Gaussian laser beam. The laser power P is constant, and the image parameter A is calculated by formula (7.4). The rest of the model parameters are as follows: $M = 1{,}024$ pixels, $N = 1{,}024$ pixels, $x_0 = 512$ pixels, $y_0 = 512$ pixels, $w_x = 2$ pixels, $\alpha = 0°$, the RMS deviation of the noise level is 2 r. u., the background illumination $C = 7$ r. u., and the size of the algorithm window is w_x. One can see from the figure that as the dimension w_y is increased,

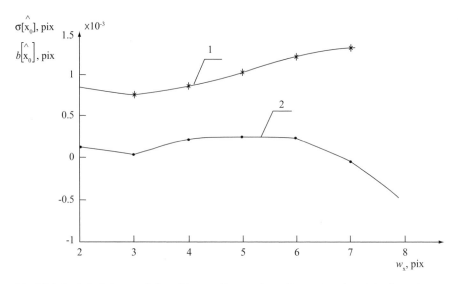

Fig. 7.26 Statistical characteristics of the coordinate estimate $\widehat{x}_0$ as a function of the dimension w_x of the Gaussian laser beam: *1*—RMS deviation of the estimate; *2*—bias of the estimate

the error of the estimate of the parameter x_0 decreases insignificantly. The random component contributes much to the total error.

Thus, the analysis of the statistical characteristics of the errors of the processing algorithm based on the method of weighting model refractograms for astigmatic laser beams allows drawing the following conclusions:

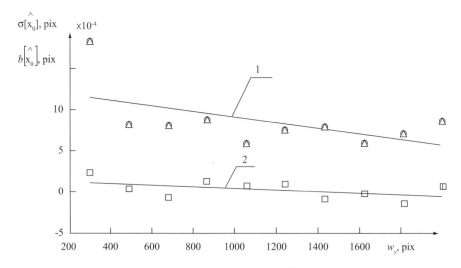

Fig. 7.27 Statistical characteristics of the coordinate estimate $\widehat{x}_0$ as a function of the dimension w_y of the Gaussian laser beam: *1*—RMS deviation of the estimate; *2*—bias of the estimate

1. To record refractograms, use should preferably be made of matrix photoreceivers with a large number of pixels (over 512×512).
2. To reduce the bias of the estimates of the parameters x_0 and α, it is expedient to reduce the inclination angle α and the dimension w_x of the Gaussian laser beam at the initial stage, when adjusting the probe radiation source and receiver.
3. The dimension w_y of the Gaussian laser beam has a weaker effect on the error of the parameter estimates than its counterpart w_x; for each size of the CCD matrix, there exists a certain range of w_y values with which the error in computing the parameter x_0 is minimal; for $M = 1{,}024$ and $N = 1{,}024$, the value of w_y should range from 800 to 1,200 pixels.

7.4.3   Numerical Refractogram Analysis Methods

In a number of cases, theoretical refractograms are computed numerically by the finite element method [13]. No parametric refractogram models figure here, and to compare between experimental and theoretical results, use can be made of the standard methods for estimating the deviation of functions [10, 14]. The theoretical relationships in such cases are specified numerically in the form of arrays of spatial coordinates of the points imaging the intersections of the geometric-optical rays of the probe laser radiation passing through the medium under study with the ground glass screen. The experimental results in the form of refractograms are processed by means of special algorithms in order to recover the contours of individual image elements, lines of maxima, distribution centers, etc. [15, 16]. The type of the line to be recovered is determined by the shape of the actual refractogram. The degree of coincidence between the theoretical and experimental results is estimated by the RMS deviation of the coordinates of the points of intersection of the geometric-optical rays with the plane of the screen from the corresponding coordinates of the points belonging to the recovered lines in the experimental images.

7.5   Digital Refractogram Recording and Processing Recommendations

The results of investigations into digital refractogram recording and processing systems allow one to give some recommendations concerning expedient ways to build such systems and draw conclusions as to the practically attainable accuracy in estimating the parameters of refractograms.

Laser refractographic systems should most advisably use special-purpose digital image recording and processing means providing for high-quality recording of refractograms and their reading in the computer memory. Such a system can be exemplified by the *Videoscan-285/B-USB* digital system considered in this chapter. The analysis of the characteristics of this system shows the importance of the proper choice of the operating conditions of the video camera, namely, operation in the linear region of the light-transfer characteristic and with an optimal relation between the camera amplification and exposure time. The necessary power of the probe laser radiation should preliminarily be determined such that the illumination of the screen should not be below the minimum permissible value, and the signal-to-noise ratio in the refractogram being processed not below the value specified to ensure the necessary accuracy in estimating the refractogram parameters. The statistical characteristics of noise in the image under processing can approximately be estimated by the method described in Sect. 7.2. Based on the a priory information on the noise characteristics, one can decide about the need to take account of the shot noise of the image being recorded, quantization noise, thermal generation noise, and electric noise pickup from the electronic circuit, and select an optimal algorithm for estimating refractogram parameters as well.

The method described in this chapter for estimating the resolving power of the recording system and the error of the estimate of the displacement of refractogram elements by means of optical test charts makes it possible to properly select the type of the objective lens for the camera and the other parameters of the optical system of the laser refractograph for the laser refractography technique to attain the necessary precision and resolution.

The analysis of the results of the development and computer-aided modeling of algorithms for the parametric estimation of laser refractograms shows that the algorithms and programs presented can be used to estimate the parameters of model refractograms for the laser plane both in the absence of the optically inhomogeneous medium under study and in its presence. The study of the characteristics of the processing algorithm based on the weighting method points to the advisability of using CCD matrices with a large number of pixels, reducing the laser plane inclination angles being recorded by properly adjusting the recording equipment, and optimizing the shape and size of the screen illumination distribution produced by the laser plane. When studying media with a radially symmetric refractive index distribution, it seems expedient to use the elliptical model of the distorted laser plane image, whose parameters provide information about the extent and character of the optical inhomogeneity of the medium under study in a convenient and compact form.

When refractograms are theoretically computed by the finite element method, the degree of coincidence between the theoretical and experimental results is estimated by the RMS deviation of the coordinates of the points of intersection of the geometric-optical rays with the plane of the screen from the corresponding coordinates of the points belonging to the lines of one or another type (contours of individual image elements, lines of maxima, distribution centers, etc.) recovered in the experimental images.

References

1. S. I. Neizvestnyi and O. Yu. Nikulin, "Charge-Coupled Devices—the Basis of the Modern Television Technology. Main Characteristics of CCDs," *Spetsialnaya Tekhnika*, No. 5, pp. 11–14 (1999).

2. E. V. Fedorova and L. A. Razumov, *Photosensitive Charge-Coupled Devices* (Moscow Power Engineering Institute Press, Moscow, 1999) [in Russian].

3. J. Dunn, *Faster Smarter Digital Video* (Microsoft Press, 2002).

4. M. Milchev, *Digital Photo Cameras* (Piter Publishing House, Saint Petersburg, 2004) [in Russian].

5. Website *http://videoscan.ru/*.

6. V. S. Ivanov, A. F. Kotyuk, A. A. Liberman, *et al.*, *Photometry and Radiometry of Optical Radiation (General Course). Book 2. Measurement of the Energetic and Space-Energetic Parameters and Characteristics of Laser Radiation (Energetic Laserometry)* (Poligraf Servis, Moscow, 2000) [in Russian].

7. I. S. Gruzman, V. S. Kirichuk, V. P. Kosykh *et al.*, *Digital Processing of Images in Informational Systems: A Tutorial* (Novosibirsk State Technical University Press, Novosibirsk, 2000) [in Russian].

8. E. V. Savchenko, L. A. Razumov, and B. S. Rinkevichyus, "Determination of the Center Coordinates of a Gaussian Beam with a Matrix Photoreceiver by the Weighting Method," *Izmeritelnaya Tekhnika*, No. 12, pp. 11–14 (2003).

9. V. A. Grechikhin and I. L. Raskovskaya, "Analysis of the characteristics of a system for digital recording of optical signals based on an array photodetector," *Measurement Techniques*, **52**(4), pp. 376–383 (2009).

10. V. S. Korolyuk, N. I. Portenko, A. V. Skorokhod, and A. F. Trubin, *Handbook of the Probability Theory and Mathematical Statistics* (Nauka, Moscow, 1985) [in Russian].

11. P. I. Rudakov and I. V. Safonov, *Processing of Signals and Images MATLAB 5.x*, Ed. by V. G. Potemkin (DIALOG-MIFI, Moscow, 2000) [in Russian].

12. H. Cramer, *Mathematical Methods of Statistics* (Almquist and Wiskell, Stockholm, 1946).

13. I. L. Raskovskaya, B. S. Rinkevichyus, and A. V. Tolkachev, "Laser Refractography of Optically Inhomogeneous Media," *Kvantovaya Elektronika*, No. 12, pp. 57–61 (2007).

14. R. Varga, *Functional Analysis and Approximation Theory in Numerical Analysis* (Society for Industrial and Applied Mathematics, New York, 1971).

15. V. A. Soifer, Ed., *Computer Image Processing Methods* (Fizmatlit, Moscow, 2001) [in Russian].

16. R. Gonzalez and R. Woods, *Digital Image Processing* (Prentice Hall, New Jersey, 2002).

Chapter 8
Laser Refractography—a Method for Quantitative Visualization of Optically Inhomogeneous Media

8.1 Principles of Quantitative Diagnostics of Inhomogeneity Profiles

The discrete and regular character of structured laser radiation makes it possible to quantitatively diagnose the media of interest on the basis of experimental refractograms. Quantitative diagnostics envisages the solution of the inverse problem and consists in the determination of the parameters of the inhomogeneity in hand (provided that it is specified by a parametric model) or the reconstruction of its profile in the form of a finite set of numbers.

To diagnose the inhomogeneities considered in the preceding chapters is essentially a tomographic problem that is generally solved using the Radon integral transformation [1]. In the case of axially symmetric distributions, the Radon transformation reduces to the Abel transformation that can be inverted analytically. The problems arising when using the Abel integral equation are due to the incorrectness of the inversion problem, for it is necessary to differentiate noisy experimental data and also overcome singularity in the integrand.

An alternative approach can involve the solution of the direct refraction problem for the family of rays forming the given structured laser radiation and calculation of the corresponding refractograms. The subsequent computer processing of experimental refractograms and their comparison with theoretical ones allows one to select such a refractive index variation law as to ensure the best fit between the theoretical and experimental data. The error of measurement depends on many factors, some of which have been considered in Chap. 7. The ultimate accuracy of gradient refractive index measurements is mainly limited by the diffraction effects whose influence can largely be eliminated by means of special refractogram processing methods (see Chap. 7).

Let us consider the consecutive stages of quantitative diagnostics [2], using as an example the analysis of refractograms for a thermal boundary layer:

1. The two-dimensional refractogram of the boundary layer is recorded with a CCD photo camera.

B. S. Rinkevichyus et al. (eds.), *Laser Refractography,*
DOI 10.1007/978-1-4419-7397-9_8, © Springer Science+Business Media, LLC 2010

2. The refractogram is processed by means of a special computer program to minimize the diffraction effects.
3. The digitized experimental refractogram is compared with a set of library refractograms computed for the given experimental setup and typical thermal layer
   profiles.
4. Based on the minimum root-mean-square (RMS) deviation criterion, the theoretical refractogram that best fits the experimental refractogram is selected.
5. The profile corresponding to the selected theoretical refractogram is taken to
   serve as the temperature profile obtained experimentally.

The suggested approach to the solution of the inverse problem of the reconstruction
of the refractive index field in transparent optical inhomogeneities makes it possible
to avoid the problems associated with the inversion of the Abel equation for noisy
experimental data. The refractography method allows reconstructing the refractive
index field of not only bulk inhomogeneities, but also of thin boundary layers, and
under substantially nonstationary conditions at that. What is more, the use of con-
stitutive equations establishing relations between the refractive index and the other
parameters of the medium under study makes it possible to diagnose the fields of
the physical parameters of the medium that influence its refractive index. The temperature field reconstruction algorithm is considered in detail in the next section.

8.2 Plane SLR Refractogram-Based Algorithm for Reconstructing the Temperature Field in a Spherical Boundary Layer

Considered in this section is the algorithm for the consecutive reconstruction of the
refractive index and temperature fields in a liquid, based on experimental refractograms of a plane structured laser radiation. Figure 8.1 presents a schematic diagram
of the experimental setup, the geometrical dimensions indicated (in millimeters)
being used in calculating the refraction of the laser plane in the inhomogeneity
under study.

As stated above, at the root of the method for reconstructing the profile of the inhomogeneity lies the comparison between the theoretical laser plane refractograms
calculated for the given parameters of the experimental setup and their experimental
counterparts that have preliminarily undergone special processing.

In Fig. 8.2, which illustrates the algorithm being considered, module *1* corresponds to the operations involved in the preliminary processing of the experimental
refractograms, including (see Fig. 8.3) the representation of the obtained images in
the form of a set of digital data, construction of the refractogram in the Matcad or
Matlab environment, and its smoothing on the basis of spline functions.

The digitized data $f(x)$ produced in module *1* enter into module *2*, both f and x
being measured in units of pixels. In order to make a comparison with the theoretical refractogram, wherein the displacement of rays is, as a rule, measured in mil-

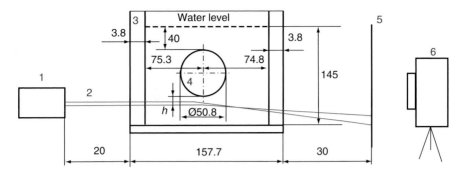

Fig. 8.1 Geometrical dimensions of the experimental setup: *1*—semiconductor laser with the laser plane forming block, *2*—laser plane, *3*—water-filled cell, *4*—metal ball, *5*—screen, *6*—CCD video camera

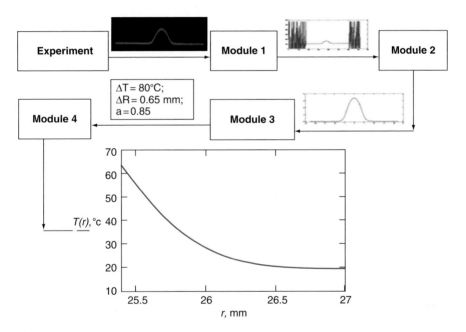

Fig. 8.2 General structure of the algorithm for reconstructing the temperature field in the boundary layer

limeters, *f* and *x* should be recalculated into millimeters as well. Prior to that, the noise present in the digital data is filtered out and the digitized experimental refractogram is displaced and turned in the proper way to make possible its comparison with the theoretical refractogram. The corrections associated with the displacement and rotation of the refractogram are necessitated by the specificities of obtaining photographic images, the selection of the perspective, and so on. After being produced at the output of module *2* the processed experimental refractogram is then completely ready for comparison with its theoretical counterpart (Fig. 8.4).

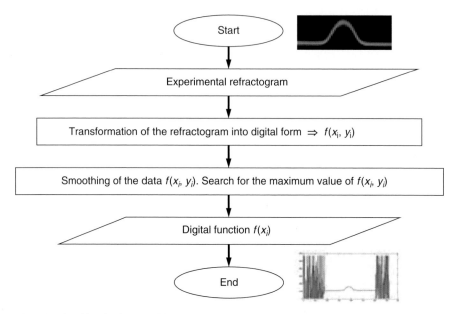

Fig. 8.3 Algorithm implemented in module *1*

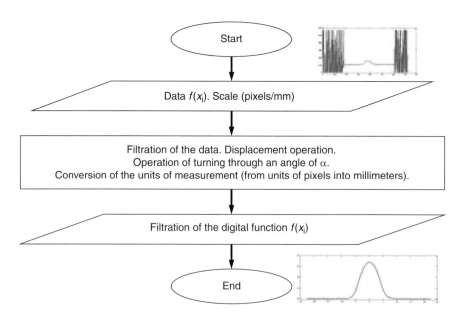

Fig. 8.4 Algorithm implemented in module *2*

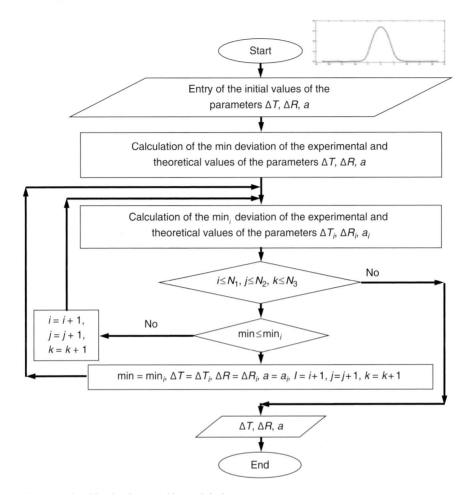

Fig. 8.5 Algorithm implemented in module *3*

Implemented in module *3* (Fig. 8.5) is the comparison of the processed experimental refractogram with the library refractograms calculated for the given parameters of the experimental setup and various inhomogeneity models. Here and elsewhere in the text, by library refractograms will be meant the refractograms of typical inhomogeneities (see Sect. 8.5). In the given case, the typical inhomogeneity is taken to be a spherical layer with a monotonic refractory index profile. The typical inhomogeneity parameters are selected to make the theoretical refractogram fit best the experimental one, as judged by the minimum RMS deviation criterion.

To the input of module *4* (Fig. 8.6) come the parameters of the inhomogeneity model chosen, which makes it possible to directly reconstruct the refractory index or temperature profile in the layer.

Fig. 8.6 Algorithm implemented in module *4*

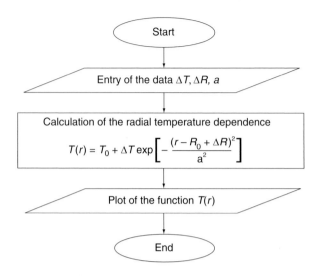

8.3 Results of the Reconstruction of the Temperature Profile of the Spherical Boundary Layer

Figure 8.7a presents the experimental refractogram (curve *1*) and the selected theoretical refractogram (curve *2*) figuring in an investigation into the process of convection in water near the surface of a hot steel ball 50.8 mm in diameter within

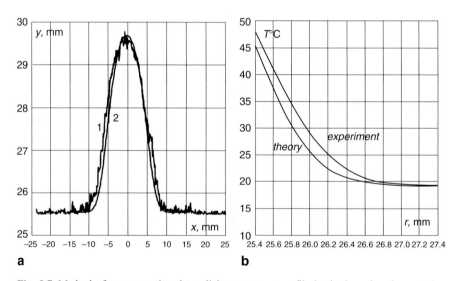

a   **b**

Fig. 8.7 Method of reconstructing the radial temperature profile in the boundary layer. **a** *1*—experimental refractogram, *2*—theoretical refractogram; **b** Experimental and theoretical radial temperature dependences

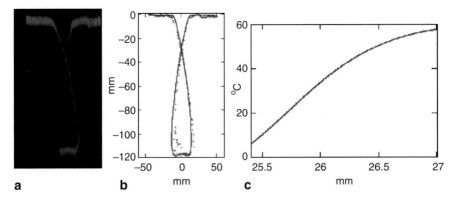

Fig. 8.8 a Experimental refractogram, **b** processed refractogram, and **c** temperature profile of the boundary layer at the surface of a ball cooled to 7°C and immersed in water with a temperature of 65°C

40 s of its immersion in water. Prior to its immersion, the ball was heated to 90°C. The center of the laser plane is located at a distance of h from the lowermost point of the ball. The thickness of the laser ball in water beneath the ball is 37 μm with respect to the $1/e$ intensity level. The refractogram was observed at a distance of 108 mm from the center of the ball. Figure 8.7b shows the radial temperature dependence in the boundary layer, reconstructed by the method described above, and its theoretical counterpart obtained on the basis of the FLUENT application program package for computing convection by the finite volume method. Comparison between the theoretical and experimental results shows a good coincidence; the disagreement does not exceed 10% and can be due to incomplete correspondence between the experiment and the model used for computation in the application program package.

Figure 8.8 demonstrates recorded and processed refractograms and the respective radial temperature distributions in the boundary layer at the surface of a ball cooled to 7°C and immersed in water with a temperature of 65°C.

Figures 8.9a–f present a series of experimentally recorded and processed refractograms and their respective radial temperature distributions $T(r)$ in the boundary layer at the surface of a hot ball (preliminarily heated to 95°C) cooling in water at consecutive instants of time t. The scheme and parameters of the experimental setup used to study free convection in liquids have been described in Chap. 5.

The plot of Fig. 8.10 shows the radial temperature distributions $T(r)$ in the boundary layer at different instants of time corresponding to the moments the refractograms were recorded. This plot illustrates the process of the cooling of the ball from the standpoint of both the variation of the ball surface temperature (the ball radius is $r = 25.4$ mm) and the qualitative and quantitative variation of the boundary layer profile with time.

At the initial instants of time (curves *1–3*) no substantial change in temperature occurs in the immediate vicinity of the ball surface, but the temperature gradient

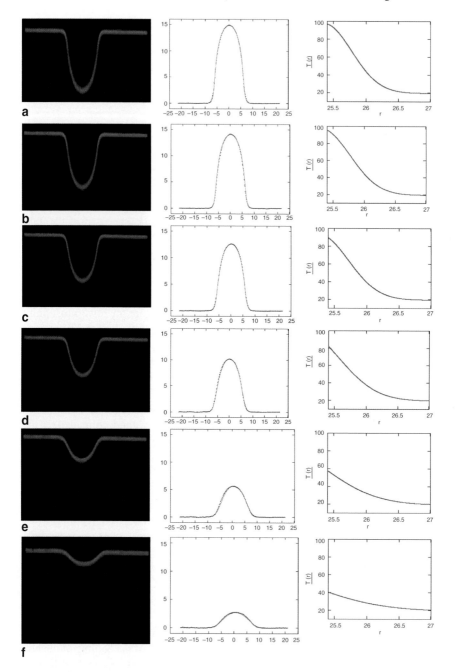

Fig. 8.9 Experimental and processed refractograms and their respective radial temperature distributions in the boundary layer at the surface of a hot ball cooling in water at consecutive instants of time. **a** $t = 0.3$ sec, **b** $t = 1$ sec, **c** $t = 4$ sec, **d** $t = 10$ sec, **e** $t = 30$ sec, **f** $t = 60$ sec

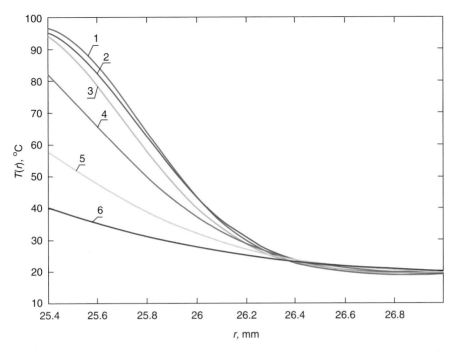

Fig. 8.10 Temperature variation in the boundary layer during the course of cooling of a hot ball in water: *1—t* = 0.3 sec, *2—t* = 1 sec, *3—t* = 4 sec, *4—t* = 10 sec, *5—t* = 30 sec, *6—t* = 60 sec

varies perceptibly from a value close to zero to some maximum value at which the surface temperature starts dropping rapidly (curve *4*). As time elapses, the temperature gradient at the ball surface drops and its cooling slows down (curves *5* and *6*). Curves *1–3* are well-approximable by a Gaussian function, while curves *4–6*, by an exponential one. The results of the thermophysical calculations of the temperature distribution in the boundary layer for a quasistationary convection process (Fig. 8.11), based on the application program package [3], show good agreement with the exponential distributions $T(r)$, but fail to agree with exponential curves *1–3*, which can be explained by the fact that the convection process is substantially nonstationary at the initial instants of time.

Thus, the laser refractography method is a fine tool for studying nonstationary processes in microscopic layers that makes it possible to reveal new effects not amenable to description by the classical methods for calculating convection processes. It is of interest to continue investigations into the relationships of the parameters of the layer profile and its variation with the physical characteristics of the surface and its microstructure.

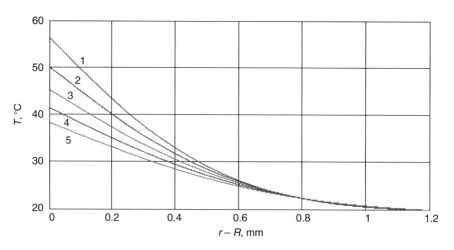

Fig. 8.11 Calculated temperature distribution in the boundary layer: 1—$t = 20$ sec, 2—$t = 30$ sec, 3—$t = 40$ sec, 4—$t = 50$ sec, 5—$t = 60$ sec

8.4 Determination of the Parameters of an Exponential Model of a Boundary Layer at the Surface of a Heated Ball

If we are given a parametric model of the inhomogeneity of interest, its quantitative diagnostics can be reduced to the determination of the model parameters. For example, when analyzing the quasistationary convection process near a heated ball in water [3], the refractive index distribution in the boundary layer can, to a good approximation, be described by the exponential relation

$$n(r) = n_0 - \Delta n e^{-\frac{(r-R)}{a}}, \qquad (8.1)$$

where n_0 is the refractive index of water in the absence of the ball, Δn is the variation of the refractive index, R is the ball radius, and a is the thickness of the boundary layer. In this model, the quantities n_0 and R are known. It is necessary to find the two unknowns—Δn and a—on the basis of measurements.

The quantity Δn can be determined from relation (8.1), provided that the radial coordinate of the turning point of the ray is equal to the ball radius, $r_t = R$; i.e., the ray touches the ball at the turning point, which corresponds to the equation

$$Rn(R) = n_0\rho_0, \qquad (8.2)$$

where ρ_0 is the distance from the center of the sphere to the top boundary of the laser plane. Experimentally this fact manifests itself as follows. As the laser plane approaches the surface of the ball, its top part first contacts the ball and then becomes screened by it, so that in the region corresponding to the rays shut out by the ball appears a "break" in the refractogram. In experiments, one should measure the quantity ρ_0 corresponding to the screening limit. Obviously, $\rho_0 = R - h$, where h is the distance from the lowermost point of the ball surface to the top boundary of the laser plane at the moment it starts to be shut out by the ball. In that case,

$$n(R) = \frac{n_0 \rho_0}{R},\tag{8.3}$$

and

$$\Delta n = n_0 - n(R) = n_0 \left(1 - \frac{\rho_0}{R}\right) = n_0 \frac{h}{R}.\tag{8.4}$$

Note that based on expression (3.17), the temperature directly on the surface of the ball can be found using formula (8.3).

To find the parameter a, one should measure the thickness Δa of the boundary layer; i.e., the distance from the surface of the ball at which the temperature of the boundary layer becomes practically equal to that of the surrounding medium. We will assume that for the exponential dependence, the relation $\Delta a = 3a$ is satisfied to a good approximation. Consider the method of processing refraction images to determine the thickness of the boundary layer at a ball [4]. To this end, use should be made of two refractometric images, one with the undistorted laser plane profile and the other with the laser plane profile distorted by the ball. Taken as an example here are the two images presented in Fig. 8.12, one obtained with no ball present and the other, in the presence of a ball heated to 70°C.

The scheme of determining the thickness of the boundary layer, given the deflection region of the laser plane, is shown in Fig. 8.13.

Based on the width Δy of the deflection region, one can determine the confines of the boundary layer and also the minimum parameter φ of the ray that no longer suffers deflection in the layer. The quantity Δa can be found from the expression

$$\Delta a = \sqrt{\Delta y^2/4 + (R + h)^2} - R.\tag{8.5}$$

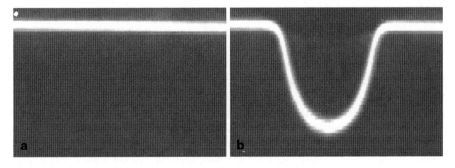

Fig. 8.12 Laser plane refractograms. **a** Original laser plane profile, **b** Distorted laser plane profile

Fig. 8.13 Determining the
thickness of the boundary
layer near a heated metal
ball: *1*—ball, *2*—laser plane
refractogram

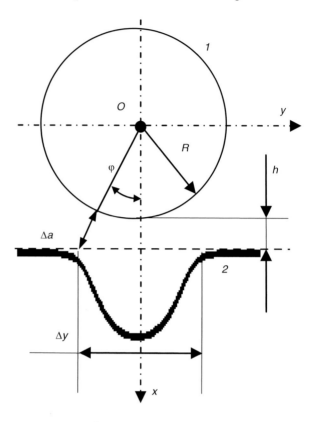

To experimentally determine the width Δy of the laser plane deflection region, one
can use the following method.

Each image in Fig. 8.12 is a matrix whose elements are the brightness values of
the picture dots. We designate the original matrix with the undistorted laser plane
trace as M_0, and that with the distorted trace as M_1. To determine the width of the laser plane deflection region, we have the difference image (matrix) M_2 wherein each
element is the magnitude of the difference between the respective elements of the
matrices M_1 and M_0: $M_{2i,j} = |M_{1i,j} - M_{0i,j}|$. The image thus obtained is presented
in Fig. 8.14a. Where the positions of the original and the distorted trace coincide,
there is no laser plane image.

Any image contains both additive and multiplicative noise. The presence of
noise makes the laser plane profile in the image lose smoothness. This circumstance
has an adverse effect on the accuracy of processing the experimental data. In the
case under consideration, the major effect on the final result is exerted by the additive noise due to the stray lighting of individual pixels, making the background
of the image not "ideally black"; i.e., in the region where there is no signal in the
difference picture, the brightness level is other than zero. For this reason, the image
should be filtered.

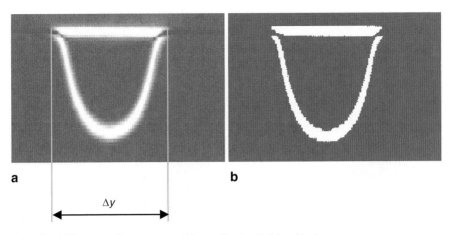

Fig. 8.14 Difference refractograms. **a** Prior to filtering, **b** After filtering

To remove the additive noise, use can be made of threshold filtering. Since the magnitude of the signal in the image is high enough, we set the filtering threshold level at half the maximum brightness. All the picture dots with a signal exceeding the threshold value assume the binary value "1" (signal is present), the rest of them assuming the value "0" (no signal present). The binary image thus obtained is presented in Fig. 8.14b.

In the general case, the threshold value can be selected from various considerations, but the lower the threshold, the higher the probability of error in the subsequent processing due to the development of a "parasitic" signal. In contrast, a high threshold level clips the signal too much.

In Fig. 8.14b, one can clearly see the borders of the laser plane distortion region. By applying this method to all the images obtained during the course of cooling of the ball, one can get the time dependences of the thermal layer parameters.

8.5   Refractogram Library Construction Principles

The quantitative diagnostics of the profile of an inhomogeneity, or, given its parametric model, of the parameters of the latter, is based on the use of a library of typical refractograms. This library is used for the preliminary identification (express diagnostics) of the type of the inhomogeneity in hand and the dynamic process being observed. Library refractograms are divided first of all into two types: theoretical and experimental. These in turn are classified according to the type of structured laser radiation, visualization method (2D, 3D, screen arrangement, etc.), inhomogeneity model, and the character of the process.

In accordance with the results of visualization (recording and processing), the starting, base refractogram is selected from the library to serve as the basis for the further refinement of the model and the process, and for the determination of the necessary

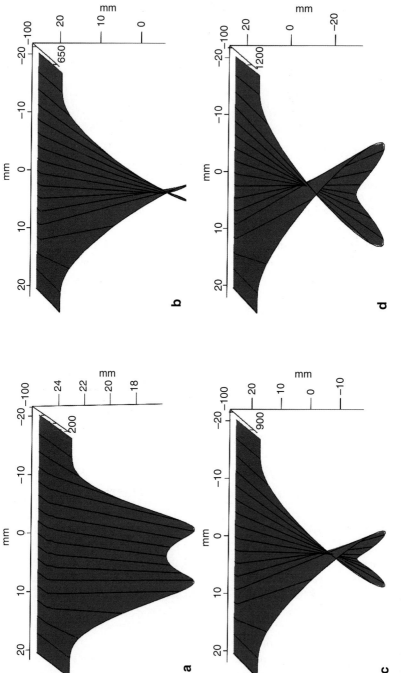

Fig. 8.15 3D refractograms for a "cold" spherical thermal layer model with the parameters: $\Delta T = 57°C$, $R = 25.4$ mm, $\Delta R = 0$ mm, $a = 2$ mm, $T_0 = 70°C$, obtained with a laser plane passing at a distance of $d = 25.5$ mm from the center of the inhomogeneity. **a** $z = 200$ mm, **b** $z = 650$ mm, **c** $z = 900$ mm, **d** $z = 1,200$ mm

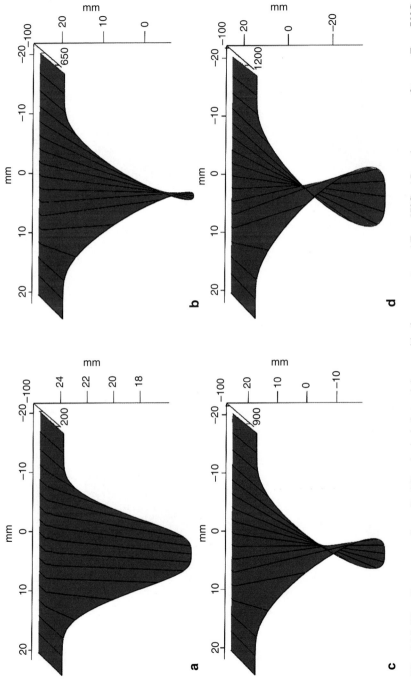

Fig. 8.16 3D refractograms for a "cold" spherical thermal layer with the parameters: $\Delta T = 57°C$, $T_0 = 70°C$, $R = 25.4$ mm, $\Delta R = 1$ mm, $a = 2$ mm, obtained with a laser plane passing at a distance of $d = 25.5$ mm from the center of the inhomogeneity. **a** $z = 200$ mm, **b** $z = 650$ mm, **c** $z = 900$ mm, **d** $z = 1,200$ mm

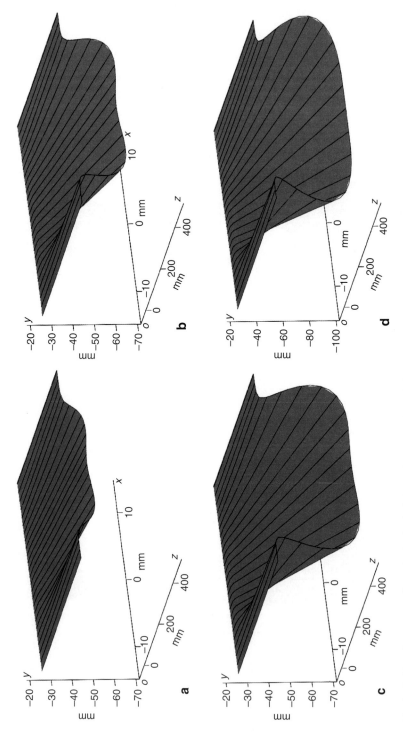

Fig. 8.17 3D refractograms for a "hot" spherical thermal layer model with the parameters: $\Delta R = 0$, $a = 1$ mm, $R = 25.4$ mm, $T_0 = 19°C$, $z = 500$ mm, obtained with a laser plane passing at a distance of $d = 25.4$ mm from the center of the inhomogeneity. **a** $\Delta T = 20°C$, **b** $\Delta T = 40°C$, **c** $\Delta T = 60°C$, **d** $\Delta T = 80°C$

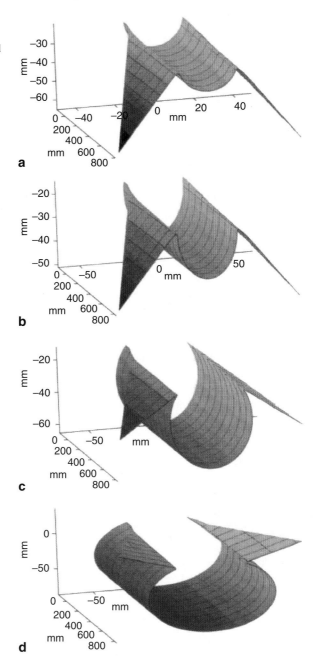

Fig. 8.18 Theoretical 3D refractograms of a cylindrical SLR for a "hot" spherical thermal layer model. **a** $r_i = 10$ mm, **b** $r_i = 20$ mm, **c** $r_i = 30$ mm, **d** $r_i = 50$ mm

quantitative characteristics. The above-described procedure allows one to state that laser refractography is a quantitative visualization method.

It should be noted that as far as the visualization method is concerned, preferable for express diagnostics are 3D refractograms, as they are more informative, because 2D refractograms for different types of inhomogeneity can be visually indiscernible, depending on the screen arrangement. In quantitative diagnostics, however, 2D refractograms prove more convenient to use.

Figures 8.15a–d and Figs. 8.16a–d present two series of library 3D refractograms for two different parameters of a "hot" spherical model according to expression (3.17). The variable parameter z in each series is the distance from the center of the inhomogeneity to the last refractogram section observed. As z varies, a loop is observed to form in the refractograms, the shape of the loop differing fundamentally between the two inhomogeneity profiles under consideration.

Figures 8.17a–d present a series of library 3D refractograms for a "hot" spherical layer model according to expression (3.17). The variable parameter in this series is the temperature of the ball surface.

Figure 8.18 presents theoretical 3D refractograms of a cylindrical SLR for a "hot" spherical thermal layer model with the parameters $\Delta R = 0$, $a = 1$ mm, $R = 25.4$ mm, $T_0 = 19°C$, and $z = 500$ mm at $x_0 = 20$ mm.

As follows from the above-described procedure for measuring temperature in boundary layers, a material merit of laser refractography is the possibilities it offers for the determination of the parameters of the medium under study and the reconstruction of inhomogenity profiles on the basis of comparison between experimental and theoretical refractograms. The error of the method depends mainly on diffraction effects and can substantially be reduced through the use of special computer image processing methods.

References

1. V. V. Pikalov and T. S. Melnikov, *Low-Temperature Plasma. Vol. 13. Plasma Tomography*, (Nauka, Siberian Publishing Company of the Russian Academy of Sciences, Novosibirsk, 1995) [in Russian].
2. I. L. Raskovskaya, B. S. Rinkevichyus, and A. V. Tolkachev, "Laser Refractography of Optically Inhomogeneous Media," *Kvantovaya Elektronika*, No. 12, p. 1176 (2007).
3. V. I. Artemov, O. A. Evtikhieva, K. M. Lipitsky *et al.*, "Investigation of a Nonstationary Temperature Field in Free Convection Conditions by the Computer-Laser Refraction Method," in *Proceedings of the 8th Scientific-Technical Conference on Optical Methods for Studying Flows*, Ed. by Yu. N. Dubnishchev and B. S. Rinkevichyus (Znak, Moscow, 2005), pp. 478–481 [in Russian].
4. K. M. Lapitskiy, "Modeling of the effect of the temperature field configuration in a fluid on the refraction of laser radiation," *Measurement Techniques*, **51**(9), pp. 1007–1011 (2008).
5. V. T. Nguyen, I. L. Raskovskaya, and B. S. Rinkevichyus, "Algorithms for quantitative diagnosis of optical heterogeneities by means of laser refractography," *Measurement Techniques*, **52**(4), pp. 368–375 (2009).
6. K. M. Lapitskii, I. L. Raskovskaya, and B. S. Rinkevichyus, "Algorithm for calculating the refraction patterns of a planar laser beam in an optically inhomogeneous medium," *Measurement Techniques*, **52**(5), pp. 494–500 (2009).

Index

B. S. Rinkevichyus et al. (eds.), *Laser Refractography,*
DOI 10.1007/978-1-4419-7397-9, © Springer Science+Business Media, LLC 2010

Printed in the United States of America